THE
WEATHER WIZARD'S
CLOUD BOOK

D1364382

YANKEE ACCENT®

23 Wianno Avenue
Osterville, Massachusetts 02655

THE
WEATHER WIZARD'S
CLOUD BOOK

How You Can Forecast the Weather Accurately and Easily by Reading the Clouds

LOUIS D. RUBIN, Sr. & JIM DUNCAN

With the assistance of Hiram J. Herbert
Introduction by Joan Rubin Schoenes

Algonquin Books of Chapel Hill · 1989

The Weather Wizard's Cloud Book is based on *Forecasting the Weather*, by Louis D. Rubin, with the assistance of Hiram J. Herbert, published in 1970.

© 1989, 1970 by Louis D. Rubin and Hiram J. Herbert. Cloud photographs © 1956, 1957, 1958, 1962, 1968, 1969 by Louis D. Rubin. Text of *The Weather Wizard's Cloud Book* © 1984 by Joan R. Schoenes and Jim Duncan.

The photographs of Hurricane Betsy, taken by the United States Air Force, and of a tornado near Rainview, Texas, are furnished through the courtesy of the National Oceanic and Atmospheric Administration.

The photograph of Mount St. Helens in eruption is furnished through the courtesy of the U.S. Geological Survey, Dept. of the Interior.

The satellite photograph of a tropical storm off the North Carolina–Virginia coast is furnished through the courtesy of the National Oceanic and Atmospheric Administration.

Photographs of weather instruments are furnished through the courtesy of the Taylor Scientific Consumer Instruments Division of Sybron Corporation, Arden, North Carolina.

Library of Congress Cataloging in Publication Data
Rubin, Louis D. (Louis Decimus), 1895–1970.
  The weather wizard's cloud book.

  Based on: Forecasting the weather/by Louis D. Rubin.
  1. Weather forecasting–Handbooks, manuals, etc. 2. Clouds–Atlases.
I. Duncan, Jim, 1956–   . II. Herbert, Hiram J. III. Rubin, Louis D. (Louis Decimus), 1895–1970. Forecasting the weather. IV. Title.
QC995.R82 1984 551.6'3 84-18478
ISBN 0-912697-10-5

Algonquin Books of Chapel Hill
P.O. Box 2225, Chapel Hill, N.C. 27515-2225
A Division of
Workman Publishing Co., Inc.
708 Broadway
New York, N.Y. 10003

CONTENTS

INTRODUCTION
THE WEATHER WIZARD

M Y FATHER, Louis D. Rubin, Sr., wrote this book for the purpose of enabling others to have as much enjoyment as he did with the weather. It is not a technical work; it does not require training in meteorology; it does not employ scientific terminology. All that is needed to use it to predict weather changes is the ability to look at cloud formations, identify the photographs in the book corresponding to what is visible in the sky, and determine the direction that the wind is blowing. It is simple, and a great deal of fun.

Throughout the state of Virginia my father was widely known as the "Weather Wizard" because of his accurate predictions. For the last twenty-five years of his life, predicting the weather was his principal activity. He was not a trained scientist. He had only a seventh grade education. But he had the basic equipment that the good scientist needs: the ability to look steadily and patiently at nature, to reason from cause to effect, and not to let automatic assumptions and prejudgments interfere with his observation of what was actually happening before his eyes. He was a remarkable man, both for what he did and what he was.

He was born in Charleston, South Carolina, in 1895, the fifth in a family of seven children. Following graduation from elementary school, he had to go to work to help support the family after his parents became ill. He found a job as stock boy in a wholesale hardware store, and soon became interested in what was still the relatively new field of electricity.

Studying at night, he taught himself to be a trained electrician. At the age of seventeen he opened a little repair and installation business. Soon he had a store of his own, selling appliances and doing electrical contracting.

By 1917 the Louis D. Rubin Electrical Company was a thriving business. When the United States of America entered the World War, he turned his store over to an associate and enlisted in the Marine Corps. After the war he returned to his business, and by the mid-1920s had built it into the leading electrical company in the South. Meanwhile he had married Jeanette Weinstein, of Richmond, Virginia, and they had three children. He was prominent in civic affairs, president of the Retail Merchants Association, and his show window displays won numerous national prizes and awards.

In late 1929, just as the Depression was getting under way, he came down with an ear infection which developed into a brain abscess. He spent much of the next several years in a hospital in Richmond, Virginia. On two occasions he was given up for dead. On the second of these, when a rabbi came to pray with him before an operation, he assured the man that within six months he would shoot and win a game of craps with him! (He did.) But when he came home to Charleston for good in late 1932, it was to find that through inept management his business had failed.

In later years he liked to say that it was while he lay day after day in a hospital bed that he began paying serious attention to the progression of the cloud patterns visible through the window. Actually, he had been interested in the weather long before that. During the 1930s and early 1940s, however, in Charleston and then in Richmond, where we moved in 1942, the weather remained a hobby. But in the late 1940s he put together a little booklet of cloud pictures, collecting his examples from magazine illustrations and advertisements. Since changes in the weather developed in regular procession, heralded always by certain sequences of clouds, anybody could match what was occurring in the sky with the photographs, determine the direction in which the cloud formations were moving, and be able to forecast what would happen next. The idea was simplicity itself—but until then no one had done it. In partnership with a local printer he produced a tiny, eight-page booklet, which he then proceeded to sell to various businesses to be used as promotional giveaways.

It was not long before he became dissatisfied with the quality of the cloud illustrations. Finding that available color photographs of clouds were few and inadequate, he began making his own photographs. Within a few years he had an extensive collection of slides of cloud formations, which were in frequent demand by weather and aeronautical bureaus, encyclopedias, textbooks, and other users. He also developed a series of cloud charts in various shapes and sizes, which were sold all over the world and with the descriptive legends translated into numerous languages.

It was in the early 1950s that my father came up with the idea that soon made his name a household word in Richmond and throughout Virginia and elsewhere. In the course of reading numerous books about the weather, he became convinced that one of the principal influences upon weather was volcanic activity. An erupting volcano belched immense quantities of volcanic ash into the sky. Such ash, traveling thereafter around the earth in the jet stream high up in the troposphere, had the effect of refracting the sun's rays and thus creating atmospheric disturbances. The theory itself was by no means original; but what my father did was to begin plotting the tropospheric movement of volcanic ash in terms of estimated quantity and rate of travel. Whenever a new eruption took place, he determined how long it would take for the ash to reach the East Coast of the United States, and for how long and at what intervals thereafter it would recur.

Making his calculations with a calendar, he produced a list of forthcoming dates on which there would be "unusual weather." Issued twice yearly, these were published in the Richmond newspapers and sent to other newspapers in Virginia over the Associated Press wire, and soon one of the favorite pastimes of Virginia residents became watching whether the "Rubin days," as the newspapers and television stations called them, would arrive as scheduled. They did—and to an amazing extent. His long-range forecasts were accurate more than ninety percent of the time.

The kind of "unusual weather" conditions, of course, varied in accordance with the season. During the summer a Rubin Day might mean only a noticeable drop in the temperature, while in winter months it could signify the arrival of a blizzard. The day before one Rubin Day, forecast for early November, he told Adam Sichol, whose gasoline service station he patronized, to lay in a supply of antifreeze, because it was very likely to snow the next day. Even though snow almost never fell in Richmond so early in the fall, Sichol did so. The next day there was a ten-inch snowfall. Other service stations were caught without antifreeze, while Adam Sichol did a landslide business.

Such feats did not go unpublicized, and for the last fifteen years of his life my father's weather prognostications were a Virginia institution. People scheduled events around them. Someone in Richmond planning an outdoor wedding would telephone my father months in advance to find out whether it would be safe to proceed on such-and-such an afternoon. The publisher of the local newspapers was once in Tampa, Florida, and was due at a board meeting in Richmond the next day, which was a Rubin Day. Bad weather was imminent along the Atlantic Coast. He called my father long-distance and asked him whether it would be safe for him to fly back to Richmond, or whether he should take the overnight train and miss part of the meeting. "If you can get out of Atlanta," my father told him, "come ahead." Finding that the Atlanta airport was still open, the newspaper publisher took the plane, and he made it safely home to Richmond well in advance of the oncoming storm. Such incidents were commonplace.

It was not only my father's skill at his predictions, but his personality, the excitement he brought to his work, the obvious pleasure he took in calling his shots and being proved right by events, often in defiance of the U.S. Weather Bureau predictions, the verve and energy with which he went about everything, that made him into so notable and endearing a presence on the Virginia scene. He was a showman, and he knew it and enjoyed it. Intuitively he knew how to give his pronouncements a Delphic quality, so that there was almost always an "if" or a riddle involved. He was good copy for the newspapers, and the television weather people were constantly checking with him. For some years he lived on Wythe Avenue in Richmond, and he was known far and wide as the "Wythe Avenue Weather Wizard." When he died suddenly, on July 30, 1970, it was front page news, and there were documentaries about him on television stations throughout the state. He would have loved it!

In the 1960s, with his booklets and charts in production, he decided to produce a full-length book about weather predicting. It would be based on his cloud photographs, and would tell about the weather from the standpoint that he had learned about it—as a layman, without scientific training. There were numerous books about the weather, but no simple, practical guide that used easily recognizable cloud photographs in color to enable anyone to make accurate forecasts. Writing, however, was not among my father's talents. So one of his sons put him in touch with Hiram J. Herbert, of Roanoke, Virginia, who was a professional author with considerable expertise in writing about science for laymen. They became devoted friends, and their collaboration resulted in a book, which they entitled *Forecasting the Weather*. A contract was signed with a publisher, and the manuscript was in the page proof stage, with my father eagerly awaiting the finished book, when he died. Published several months later, the book received little attention; when the original press run was exhausted, it went out of print. Meanwhile his cloud charts and booklets were marketed by Cloud Charts, Inc., and continued to be widely used.

It is this book, revised and rewritten, that provides the basis for *The Weather Wizard's Cloud Book*. In the fourteen years since the first version was published, there have been certain changes in weather knowledge and terminology. Meteorologists now place much more importance on the effects of volcanic activity on local weather conditions than they did when my father first began using that activity to make his long-range predictions. The eruption of Mount St. Helens, squarely within the continental United States, clearly had striking effects on customary weather patterns. So it seemed appropriate to bring the book up to date, and add a section on volcanoes, even though the basic method of weather forecasting, the information about weather patterns, and the array of photographs are as sound as ever.

Hiram J. Herbert, who had worked with my father on the book, passed away in 1982. To revise, rewrite, and otherwise bring the book up to date we were fortunate to secure the co-authorship of Jim Duncan, meteorologist for WWBT-TV, Richmond, an accomplished student of weather prognosticating himself, who has written and lectured extensively on the subject.

Here, then, is *The Weather Wizard's Cloud Book*, a book about weather forecasting, designed to offer readers some of the joy and excitement that my father found in observing the clouds and predicting what the weather will be. It is offered in the same spirit of high-heartedness and devotion that he brought to everything he did. It is his book, and I hope you will like it.

JOAN RUBIN SCHOENES

Richmond, Virginia
January 30, 1984

WILL TOMORROW BE FAIR? Rainy? Hot? Cold . . . Cool? Clear? Cloudy? Windy? Calm? In "language" not difficult to read, the vast face of the sky tells us what the weather is going to be.

Each cloud formation we see contains a weather message, an advance notice. These signs in the sky are easy for us to read. The constantly changing cloud formations are Nature's own weather prophets, accurately forecasting the weather variations minutes, hours, and even days ahead. While the factors that produce our weather are complicated, they are not at all mysterious. Once we learn to recognize the sky message, easily interpreted, the weather that follows comes as no surprise to us.

Weather moves with the winds, generally from west to east. Any significant change in existing weather is usually accompanied by a change in wind direction: wind change, weather change. This is true the world around, because cloud formations are the same everywhere.

Once we learn to read the face of the sky we can know the kind of weather that is in store for us, and how long we must wait for it to happen. The purpose of this book is to show graphically—and simply—how we can read the sky for the weather news.

The natural-color sky and cloud photographs in this book should enable anyone to predict the kind of weather to expect from day to day with reasonable accuracy. Usually, a single

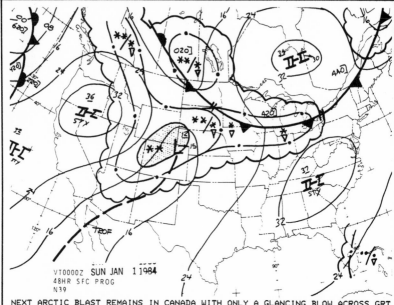

NEXT ARCTIC BLAST REMAINS IN CANADA WITH ONLY A GLANCING BLOW ACROSS GRT LKS BUT THICKNESS VALUES ACTUALLY HIGHER THAN THOSE PRESENTLY.
AS FOR REST OF COUNTRY THINGS SETTLE DOWN AND SLOWLY MODERATE UNDER EX-PANSIVE SURFACE RIDGE. ONLY PRECIP PROBLEM LEFT IS IN FLORIDA WHERE SAT DATA SHOWS HIGH AND MID LEVEL MOISTURE PERSISTING OVER ERN GULF SUGGEST-ING SOME POSSIBLE FRONTAL WAVE ACTION. HIGH POPS HINT AT SUCH BY 24 HRS.

IN CLOSING THE TIME HAS COME TO RING OUT THE OLD AND BRING IN THE NEW AS ANOTHER YEAR IS ABOUT TO BE TACKED ON TO THE CALENDAR. THUS YOU ALL HAVE A HAPPY AND HEARTY NEW YEAR. SEE YA NEXT YEAR................POOLE

Fig. 1 Typical weather map furnished by the National Weather Service. *Courtesy National Oceanic and Atmospheric Administration.*

picture tells us all we need to know. Sometimes a series of cloud formations passing in review is necessary for us to read the weather outlook.

Nothing affects each person, wherever he may live, as completely as does the weather. Weather conditions are fundamental in the production of our foods, in scheduling our recreation, in what we wear, in the planning of work, in types of housing, in the progress or slowdown of business, industry, and transportation, in the state of our health, in our successes and our failures. The psychological effects of weather on the human temperament are little short of astounding.

The behavior of the atmosphere, which includes the cloud formations, directly affects the behavior of everything, animate or inanimate, on the earth. It supports life and also brings death. It brings joy and also grief. The Great Ice Age was a manifestation of weather behavior, as was the great Biblical Flood. The climatic zones—such as polar, torrid, and temperate—that encircle the earth are the exclusive works of weather in orderly, departmentalized control of our vast sphere, the angle of the sun and the air currents masterminding the total pattern.

The face of the sky is our constant, dependable barometer. Its many "expressions" tell us *what* the weather is going to be, *when* it will change, and to *what* degree. The ancients were well versed in reading the face of the sky. They were skilled in accurately interpreting what they saw written in the heavens. The form and composition and color of the clouds and the direction of the winds that moved them told the story of what to expect of the weather. Through the centuries this art became lost as the advancements of civilization produced more creature comforts.

The face of the sky can be bright and cheerful. It can be angry, or stormy, or moody, or unsettled, or threatening, clearly conveying to us its "intention," as if "telling" us exactly what is to come. It is there for us to see—fair weather, wet weather, stormy weather, cold or warm weather. The sky "tells" us when the thunderstorms are due.

When we travel from country to country, even from state to state, we see things that differ greatly in form and in substance, often to extremes. The Blue Ridge Mountains are in sharp contrast with the Himalayas; the Alps are unlike the Ozarks. Arabia bears little resemblance to New England, nor does the land of the Eskimo hold much in common with the West Indies. Species of animal and plant life contrast vividly, country by country. Soil types differ. The people are different. From place to place, almost everything on the earth is different.

But the clouds, *the many faces of the sky*, are the same the world around. The cumulus cloud, that detached spectacular one so beautifully thick and snowy, appearing as a dome or a towering mountain, is the same rising over Indiana as the cumulus rising over China or Chile, over Spain or Maine. Similar conditions cause the formation of cumulus clouds (and all other clouds) wherever they appear. In each of these widely separated places, people who possess the knowledge of *why a certain cloud is present*, and what the direction of the prevailing wind can produce, can each know what the weather is going to be, and *when*! However, it is important to remember that below the equator winds move in the *opposite* direction from those above the equator.

No weather predicting system is one hundred percent accurate. In forecasting the weather by the use of the most expensive scientific instruments, or by a reliance upon the clouds, cloud colors, and the winds—the final forecast depends upon the individual.

In order to become adequately qualified to forecast our "own" weather, we must have a clear comprehension of the major factors that make the weather. Patience and observation are the keys that unlock the secrets of the clouds.

OUR ATMOSPHERE

First, exactly what is the *atmosphere*? When we understand the atmosphere, we understand the weather better.

The sun, although approximately 93 million miles from the earth, controls life and weather on the earth. Without the heat energy from the sun interacting with the earth's moisture sources, there would be no weather as we know it. There would be no clouds, no rain or snow, and probably no life.

Our weather is contained in a comparatively small area. From what we know, most weather is confined within the first five to ten miles of our atmosphere.

We live in what is called the *troposphere*; the part of the atmosphere closest to the earth, which ranges to about six miles above the surface at each pole, and ten or twelve miles over the equator. Within this atmospheric layer dwell the clouds that tell us the weather story.

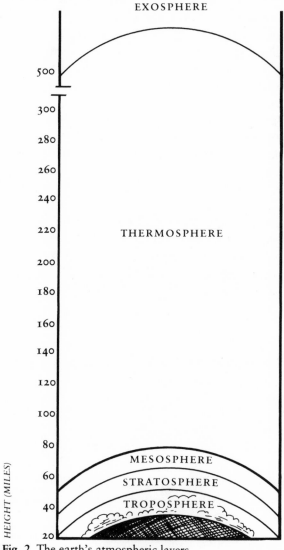

Fig. 2 The earth's atmospheric layers.

The second layer of the atmosphere is the practically weatherless *stratosphere*—extending about thirty miles above the troposphere.

The next layers are the *mesosphere* and *thermosphere*. Bands of ionized particles in the thermosphere can reflect radio waves from the earth. This explains why at night you will often hear radio stations from hundreds or even thousands of miles away.

Finally, there is the *exosphere*—starting at about 500 miles above the earth's surface, and reaching into space. This is the so-called "outer limit" of the earth's atmosphere.

All weather begins and is contained in the troposphere, the earth-circling "ocean" of air with the earth as its floor. When the actual oceans become stormy, their floors are disturbed by the turbulence of the waters. Currents that move in the ocean waters influence all life within.

The face of the earth—the floor of the atmospheric ocean of air—becomes disturbed or calm, warm or cool, wet or dry, because of the behavior of weather systems within the troposphere. Prevailing wind currents that flow through the troposphere steer and develop the highs and lows—the weather

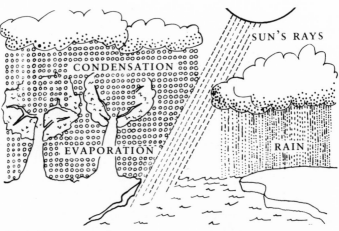

Fig. 3 The water cycle.

"makers" that change our weather from clear to stormy, and back again.

Water is drawn from the earth's surface, only to be recycled into clouds and precipitation, so that it can once again return to the ground. This cycle is continuous, and provides a vital redistribution of the earth's water, so that all parts of the world can receive their nourishment. Without water, there would be no weather as we now know it.

2 VISIBLE WEATHER, AND WHY

PRECIPITATION

PRECIPITATION is any moisture that falls from the clouds
to the ground—drizzle, rain, ice (hail, sleet), or snow. Always the atmosphere contains varying amounts of water vapor,
formed by evaporation of water on the earth. This evaporating moisture comes from moist land, the rivers and streams,
lakes, and the oceans and seas, and from vegetation such as
the grasses, shrubs, and trees. Trees, especially, liberate enormous quantities of moisture into the air.

Clouds and precipitation form when this gaseous water vapor changes back to liquid water droplets or ice crystals. For
this to happen, the air must be *cooled* to its point of saturation. It can either be cooled directly—such as by moving over
a cold body of water—or can cool by *rising* up higher into the
atmosphere.

In fact, *rising air motion* is the key mechanism in most
cloud and precipitation formation. Air can either be *forced* to
rise, such as winds blowing against a mountain or warm air
"riding" up and over a colder air mass (warm front), or can
rise as a result of temperature contrasts—when warm air near
the earth's surface rises because it is *lighter* than surrounding
air. (Just like a hot air balloon!)

As air rises, it expands and cools, and since cool air can

hold less moisture than warmer air, some of the water vapor in the rising air condenses, and turns to liquid water or ice crystals. These water droplets and ice crystals are very tiny, and stay suspended in the air to form a cloud. Eventually, the cloud droplets will grow in size, and become heavy enough to succumb to gravity, and then return to the earth's surface as some type of precipitation. The exact form that the precipitation takes when it reaches the ground depends upon the temperature conditions it encounters on its trip from cloud to ground.

Actually, the change from water vapor to liquid water droplets or ice crystals is very dependent on the amount of impurities in the air. The water vapor attaches to tiny particles of dust and other microscopically small particles, such as sea salts and acids. Thus each drop of water or bit of snow, sleet, or hail that we see contains thousands of these particles, so small they cannot be seen by the naked eye.

SNOW

The most dramatic and beautiful form of precipitation is snow, the transformer of landscapes. A product of ice clouds, snow is water in solid form. It begins as a tiny bit of ice, floating, rising, and falling in the free air. As it moves, it grows in size. The snow that can blanket the earth in quiet beauty or halt the movement of a nation of men and industry can find its birth in the altocumulus, a middle water and ice cloud.

Snowflakes form when water vapor changes directly to ice, rather than first to water droplets and then to ice. In scientific jargon, this is called *sublimation*, and will happen in a cloud only when the temperature is below freezing. The vast variety of intricate and beautiful patterns of snowflakes are infinite in number, but always hexagonal in shape.

Snow can be a dry, fine substance called "diamond dust," requiring many hours to fall 1,000 feet, or fluffy, cottony, moisture-laden flakes as much as an inch in diameter, that make the same journey within ten minutes.

Never is it too cold to snow. The colder the air, the drier and lighter the snow.

The wind direction that accompanies snow tells us much more of what to expect of the weather. North of the equator, snow that comes with the wind blowing from the northwest to the north is usually followed by colder air. Such storms may be violent, but usually they are of shorter duration than storms which come with winds from the north and northeast to the south. The same conditions apply to rain. Sometimes, the snow that comes with winds from north, northeast, and east is mixed with some sleet and rain, and the snowflakes are usually smaller. When snow rides on the winds from the northeast to the south, the storms that may last for several days are produced.

Snowfall can be estimated on a scale: ten inches of snow equals one inch of rain. This is an arbitrary comparison, since the relation of snow to moisture is influenced by how damp or how dry it is.

HAIL

Hail is mostly a product of cold-front storms, and sometimes of local storms. Despite its icy form, it occurs primarily during the hottest months of the year.

Hail is the "trapeze" performer in the huge weather tent, swinging in many directions before it plummets to the earth-net. In a violent thunderstorm, there is a strong updraft of wind. (This is why thunderstorms can be dangerous for aircraft.) The creation of hail begins when a drop of rain, or a snow pellet (graupel or soft hail), is drawn upward by the wind. As it rushes higher, it freezes. Then it falls again, but the strong updraft catches it and carries it higher and higher. This process of dropping and rising happens over and over again, bringing it into repeated collision with other supercooled rain droplets.

Thus the hail "stone" grows larger, sometimes as large as a baseball. When it becomes so heavy that the updrafts no longer can move it upward or support it, the hail "stone" falls to the earth. When driven by a strong wind, hail strikes with considerable, sometimes destructive, force.

On examining hail we will find it to be composed of layers of whitish, opaque ice. When we slice a hail "stone" in half we can see the stages of its formation, layer by layer, as it was swept into the freezing air, collected another coating of moisture, fell again, and was hurled upward again for another freezing in the upper levels of the cloud.

Most hailstorms occur between 12:30 P.M. and 8:30 P.M., and the majority of those between 3:30 P.M. and 5:30 P.M. This is because our weather is generally hotter in the afternoons than in the mornings. Hot and humid weather is an important ingredient in the production of a thunderstorm. Hail, being a part of a thunderstorm, also favors warm weather for its formation.

SLEET

Sleet is "grains of ice," ice pellets, or frozen raindrops. It forms when raindrops from the upper air fall through a layer of air which is colder than 32 degrees F. Grains of sleet fall calmly, as opposed to the turbulent journey of hail.

This form of precipitation reaches the earth in the form of small bits of ice.

Often, sleet falls in such small particles that it melts on contact with the earth, unless the earth is frozen or the surface area is freezing cold. When the ground is cold, the sleet instantly forms a pebbly crust or glaze.

FOG

Fog, dews, and frost are *not* forms of precipitation. Rather, fog may be described as a cloud in contact with the ground. It is the result of condensation that occurs in the air at, or very close to, the earth's surface. There is no essential difference between a fog and a cloud. We have water fog (water droplets), and ice fog (floating ice crystals).

The fog we see most often is the ground, or radiation fog,

usually occurring at morning when the air is cool. It can be a nuisance, even a hazard; but it usually dissipates or evaporates in a few hours.

Ground fog occurring in the morning usually assures us of no rain for that day. Ground fog occurs primarily on clear, windless nights, when heat built up at the earth's surface during the day radiates skyward with ease at night. As the earth cools, the air immediately above it cools, and eventually becomes saturated. Water vapor in the air condenses—and turns to tiny liquid droplets. Similar conditions cause formation of dew, and also frost when the temperature on the ground is below freezing.

Advection fog is produced when moist air moves over colder land or water, and is most common along the seashores, on oceans and large lakes. Duration of advection fog depends upon the force and direction of the winds. This is the type of fog that is especially hazardous to shipping and to aviation.

Another type is precipitation fog, which is more like a low stratiform cloud formation and has the appearance of a continuous drizzle. Its causes are similar to those which produce rain. Appearing usually in the spring and fall, this type of fog may last for days.

Though early morning fog is generally a sign of no rain for that day, to assure a rainless day, the previous night must have been clear. This rule does not apply to night fog.

WINDS

Wind is air in motion. It can be a gentle breeze moving horizontally over the land, a hurricane rushing at 74 miles per hour or faster, or a raging tornado dipping its tail possessed of almost incredible destructive power.

While the cloud formations are the same the world over, winds vary. Their directions and speed can be affected locally by mountains, lakes, and rivers.

On the larger scale, though, *prevailing* winds are the same at certain latitudes. For example, the *Prevailing Westerlies* dominate midlatitude regions—such as the United States.

Our general wind circulations are the result of uneven heating of the earth by the sun. Air is heated much more intensely at the equator than at the poles. Thus, the warm, light air rises in the tropics, while at the same time the cold, heavy air sinks at the poles. The rising tropical air moves northward before sinking again at around 30° latitude, and then moves back south. The cold polar air warms up some in its trek south from the top of the world, and starts to rise skyward at about 60° latitude, only to return in its poleward journey.

Between these two wind circulation "cells" is a third one which prevails at midlatitudes. The earth's rotation "pulls" the winds from all three cells to set up the prevailing wind patterns that we see in different parts of the world; in our hemisphere, the *Polar Easterlies* in northern latitudes, the *Prevailing Westerlies* in middle latitudes, and the *Trade Winds* in southern latitudes.

While winds such as the Trade Winds are fairly constant, our Prevailing Westerlies are not, and in fact they move in various cycles throughout the year. These are the winds that create our forever changing weather conditions. *If there were no winds, our weather would always be the same.*

Here are a few forecasting tips based solely on the direction of the wind.

When the winds shift rapidly from the northwest to the northeast, or from the southeast to the southwest, back and forth during clear skies or unsettled weather, there will be a change in the existing weather. In such cases, however, there is no way to determine how soon this change will come.

If the wind shifts from the southeast back and forth to the north-northwest during rain, the weather soon will clear.

If the wind shifts from the southeast to the southwest and the skies are clear, chances are good that some precipitation will soon occur.

In general, fair weather comes when the wind is blowing from the southwest, west, northwest to north, clockwise (or forward).

Unsettled weather usually comes with winds blowing from the north-northeast, east to south.

As a general rule of thumb, remember that any change in surface wind direction usually foretells a change in the weather. (All caption forecasts appearing under pictures in this book are based on surface wind conditions.)

3 THE CLOUDS—Guide to the Weather

A CLOUD is a visible composition of minute particles of dust, other infinitely tiny minerals, water, and in some formations, ice. There are water clouds and ice clouds.

Types of clouds are decided by color, size, the distribution of particles, and elevation. The type also depends upon the color and the intensity of the light from the sun and reflected light; on dimensions, shape, composition; and, finally, on the luminosity and colors *as seen from the point of observation.*

Luminosity of clouds is determined by the amount of light from the sun, light reflected by the constituent particles, and light reflected from the earth's surface. Haze affecting the colors also must be determined. At night it is difficult to determine the type of cloud when the moon is less than one-half or when it is behind high, thin clouds.

The clouds and the skies receive their light and much of their color from the diffused light rays of the sun, mixed with moisture and atmospheric dust.

The cloud pictures on following pages will serve as dependable guides in determining the clouds existing at the time we attempt to foretell the weather from them.

The information under the cloud pictures includes, in some instances, subtitles agreed upon by the World Meteorological Organization in Geneva, an arrangement which makes the language of cloud identification and meaning uniform and readily understandable in all countries.

It is no more difficult to learn cloud names than it is to learn the names of our neighbors and members of their families and how they behave! Each has a common surname, yet each differs in physical characteristics—such as complexion, hair color or texture—that distinguish them as different, yet of the same family.

Most important for weather forecasting is to be able to recognize the various clouds as they affect the weather. Anyone who learns the four basic groups can make day-to-day forecasts in his locality with reasonable accuracy.

CIRRO or CIRRUS denotes high clouds. These are ice clouds, occurring at altitudes ranging from 20,000 to 40,000 feet.
ALTO denotes middle clouds, ranging from 6,500 to 20,000 feet.
CUMULUS appears as a "heap." The "wool pack" cloud is typical.
STRATUS appears as a "layer."

There are three important factors in forecasting weather: the sequence of cloud formations, their direction of movement, and surface wind direction. This is because changeable weather is often associated with warm or cold fronts. Slow-moving warm fronts give several days' warning of pending rain of long duration by cloud changes. Rapidly moving cold fronts bring showers of short duration.

Clouds are given subtitles for more completely describing the physical variations of their formations. Also, clouds have been classified by families, as a convenient method of identification, just as we have the cat family in which we find related members distinguished by physical differences, such as the lion with his mane, the tiger with his stripes, the leopard with his spots.

Consequently, each cloud formation bears a definite Latin name because of its particular features—such as *floccus* (woolly), *uncinus* (hook-shaped). The name cirrus cumulo-nimbo-genitus sounds technical, yet it is self-interpreting and appropriate. Cirrus clouds are separate clouds of white filaments and white bands, silky, hairlike. Cumulonimbus clouds are heavy, dark, towering, sometimes shaped like an anvil or vast plumes. *Genitus*, a Latin word, like *genera*, refers to the birth, or beginning, of the cloud formation. *Genitus,* used with the other names—cirrus cumulonimbo—indicates the clouds are forming and growing from the "mother clouds" into various formations and shapes.

Cirrus cirrostratomutatus indicates, from the term *mutatus,* meaning mutation or change, that this cloud is undergoing changes with family characteristics into other forms.

With a little careful study the amateur can grasp these interpretations readily and derive more satisfaction from the ability to foretell the weather.

While knowing the names of all clouds is not necessary to weather forecasting, it is well to become familiar with the ten types of clouds which *do* affect weather. These are the main groups, with abbreviations used on weather maps, and the common subtitles:

CIRRUS (Ci.)	fibratus, uncinus, spissatus, castellanus, floccus.
CIRROCUMULUS (Cc.)	castellanus, floccus, stratiformis, lenticularis.
CIRROSTRATUS (Cs.)	fibratus, nebulosus.
ALTOCUMULUS (Ac.)	castellanus, floccus, stratiformis, lenticularis.
ALTOSTRATUS (As.)	altostratus.
NIMBOSTRATUS (Ns.)	nimbostratus.
STRATOCUMULUS (Sc.)	castellanus, stratiformis, lenticularis.
STRATUS (St.)	nebulosus, fractus.
CUMULUS (Cu.)	humulis, fractus, mediocris, congestus.
CUMULONIMBUS (Cb.)	calvus, capillatus.

Other subtitle references, such as translucidus, perlucidus, and opacus, describe the luminosity of clouds (as in the pictures that follow), where the main titles are alike but the descriptions vary.

The sequence of clouds not only is uncomplicated but is fascinating to observe; we watch the coming weather unfold before our eyes! The sequence is: first, high white ice clouds;

then darkening to become low water clouds; next comes the precipitation; and finally the skies clearing of clouds. Then the cycle begins again.

Also, code numbers have been assigned to clouds and are universal in use. These codes are a kind of "shorthand" which quickly describes, or identifies, them. "H" indicates the *high*, or cirrus, clouds, at 20,000 feet. The added number—such as H1, H2—means a variation of cirrus. "M" applies to the *middle* clouds, at 6,500 to 20,000 feet; "L" to the *lower* clouds, which form near the earth but can range, during a storm, to 25,000 feet.

Cirrus clouds are detached, delicate, fibrous, with white filaments and white bands (Plate 1–A). They have a silky, hairlike appearance, tufts, lines across a clear sky, and are sometimes called "mares' tails," "painters' brushes," "hen feathers," or "spider webs."

Cirrus clouds can be distinguished from cirrocumulus (Plate 2–A) by their silky appearance and absence of small globules which look like grains of sand or ripples at the seashore. They are distinguished from cirrostratus (Plate 3–B) by the difference in shape and thinness of their horizontal bands. Sometimes, when they are low on the horizon, cirrus clouds can be mistaken for cirrostratus clouds. Cirrus clouds are composed of ice particles.

Cirrocumulus clouds are ice clouds in sheets or layers of small globular clouds, ripples like sand, or small elements looking like fish scales, and without shading. Often, they give the appearance of a "mackerel sky" (Plate 2–A).

Cirrostratus clouds are transparent, white, and fibrous in appearance. They are composed of ice particles. They may be mistaken for stratus (Plate 2–C). These clouds give a milky look to the sky, not dense enough to blur or diffuse the outline of the sun or the moon, but occasionally to give it a halo (Plate 3–D).

Altocumulus clouds are white, although there may be some light-gray parts in flattened layers, masses, or rolls; sometimes they appear in waves, their edges touching. They may be fibrous or diffuse, but mostly in small elements. Altocumulus

PLATE I

Cloud Varieties

A—CIRRUS fibratus
Ice clouds. Good weather if
winds from W NW to N. Precipitation likely within 20 to 30
hours if winds steady from NE E
to S.

**B—ALTOCUMULUS
translucidus**
Water and ice clouds. Some precipitation likely within 15 to 20
hours if wind is steady NE to S.
Other winds bring overcast sky.

**C—STRATOCUMULUS
stratiformis**
Immediate threatener of bad
weather from a sprinkle to heavy
precipitation. If at head of cold
front, gusty winds or thundershower.

PLATE 2

Cloud Varieties *(Continued)*

A—CIRROCUMULUS

Ice clouds. Precipitation likely within 15 to 20 hours if wind NE to S. Early summer A.M. often afternoon thundershowers. Other winds overcast.

B—CUMULUS humulis

Fair weather clouds if they show no vertical development. Can build up and develop into cumulus congestus or cumulonimbus clouds.

C—STRATUS (Note tower fading in cloud)

Winds from NE to S may bring heavy precipitation. Other winds bring only light drizzle or an overcast sky.

PLATE 3

A—CUMULUS fractus

Fair weather clouds broken up by strong winds. No precipitation unless winds steady NE to S, then cumulus fractus of bad weather.

B—CIRROSTRATUS covering sky

Precipitation likely within 15 to 25 hours if winds steady from NE E to S, or sooner if winds SE to S. Other winds bring overcast sky.

C—NIMBOSTRATUS

Rain or snow clouds. Precipitation of long duration if winds NE to S, or shorter duration if winds are from SW W to N.

D—CIRROSTRATUS with halo

Precipitation likely within 15 to 24 hours if wind steady from NE to S. Prismatic effect of sun or moon through ice crystals causes halo.

PLATE 4

A—CIRRUS fibratus

Ice clouds. Good weather if winds from W NW to N. Precipitation likely within 20 to 30 hours if winds steady from NE E to S.

B—CIRRUS spissatus

Ice clouds. Good weather if winds from W NW to N. Precipitation likely within 20 to 30 hours if winds steady from NE E to S.

C—CIRRUS uncinus spreading over sky

High ice clouds. Good weather if winds from W NW to N. Precipitation likely within 20 to 30 hours if winds steady from NE E to S.

D—CIRRUS above 45°

High ice clouds. Good weather if winds from W NW to N. Precipitation likely within 20 to 30 hours if winds steady from NE E to S.

E—CIRROSTRATUS covering sky

Precipitation likely within 15 to 25 hours if winds steady from NE to S, or sooner if winds SE to S. Other winds bring overcast sky.

F—CIRROSTRATUS covering sky

Ice clouds cover sky. Precipitation likely within 15 to 25 hours if winds steady from NE E to S. Other winds bring overcast sky.

G—CIRROCUMULUS

Ice clouds. Precipitation likely within 15 to 20 hours if wind NE to S. Early summer A.M. often afternoon thundershowers. Other winds overcast.

PLATE 5

A—CIRROCUMULUS
Ice clouds. Precipitation likely within 15 to 20 hours if wind NE to S. Early summer A.M. often afternoon thundershowers. Other winds overcast.

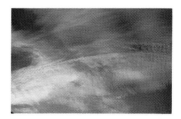

B—CIRROCUMULUS
Ice clouds. Precipitation likely within 15 to 20 hours if wind NE to S. Early summer A.M. often afternoon thundershowers. Other winds overcast sky.

C—CIRROSTRATUS with halo
Precipitation likely within 15 to 24 hours if wind steady from NE to S. Prismatic effect of sun or moon through ice crystals cause halo.

D—ALTOSTRATUS translucidus
Precipitation likely within 10 to 15 hours if winds steady NE to S. Sun appears to be behind frosted glass. Other winds bring overcast sky.

E—ALTOCUMULUS translucidus
Water and ice clouds. Some precipitation likely within 15 to 20 hours if wind is steady NE to S. Other winds bring overcast sky.

F—ALTOCUMULUS undulatus
Precipitation likely within 15 to 20 hours if wind is steady NE to S. Other winds bring overcast sky. Mostly threatens precipitation.

G—ALTOCUMULUS perlucidus
Water and ice clouds. Some precipitation likely within 15 to 20 hours if wind steady from NE to S. Other winds bring overcast sky.

PLATE 6

A—ALTOSTRATUS translucidus
Precipitation likely in 10 to 15 hours if winds steady NE to S. Sun appears to be behind frosted glass. Other winds bring overcast sky.

B—ALTOSTRATUS translucidus
Precipitation likely within 10 to 20 hours if wind is steady from NE to S. Sun appears behind frosted glass. Other winds overcast sky.

C—CUMULUS humilis
Fair weather clouds if they show no vertical development. Can build up and develop into cumulus congestus or cumulonimbus clouds.

D—CUMULUS with vertical growth
Fair weather clouds building up vertically often indicate cumulus congestus and cumulonimbus which bring afternoon showers.

E—CUMULUS congestus
If clouds form from SW to NW precipitation with gusty winds and thunderstorms or only wind squalls are likely within 5 to 10 hours.

F—CUMULUS congestus
Cloud tops forming altocumulus cumulogenitus usually bring precipitation in 5 to 10 hours, gusty winds, thundershowers, wind squalls.

G—CUMULUS congestus
If clouds form from SW to NW precipitation with gusty winds and thunderstorms or only wind squalls are likely within 5 to 10 hours.

PLATE 7

A—CUMULONIMBUS
Precipitation likely and soon coming usually from SW W to N. Distant clouds often show an anvil-shaped cirroform cap.

B—CUMULONIMBUS capillatus
Precipitation likely and soon coming usually from SW W to N. Distant clouds often show an anvil-shaped cirroform cap.

C—CUMULONIMBUS calvus
Precipitation likely and soon coming usually from SW W to N. Distant clouds often show an anvil-shaped cirroform cap.

D—STRATOCUMULUS stratiformis
Immediate threatener of bad weather from a sprinkle to heavy precipitation. If at head of cold front, gusty winds or thundershower.

E—STRATOCUMULUS opacus
Immediate threatener of bad weather from a sprinkle to heavy precipitation. If at head of cold front, gusty winds or thunderstorms.

F—STRATUS (Note tower fading in cloud)
Winds from NE to S may bring heavy precipitation. Other winds bring only light drizzle or an overcast sky.

G—CUMULONIMBUS mamma
Seldom seen low to middle clouds. Associated with severe wind squalls, hail, heavy precipitation, tornadoes, and thunderstorms.

PLATE 8

A—NIMBOSTRATUS
Rain or snow clouds. Precipitation of long duration if winds NE to S, or shorter duration if winds are from SW W to N.

B—NIMBOSTRATUS
Rain or snow clouds. Precipitation of long duration if winds NE to S, or shorter duration if winds are from SW W to N.

C—CUMULUS congestus with fracto stratus of bad weather
Precipitation of long duration if winds from NE to S, or shorter duration with other winds.

D—CUMULUS fractus
Fair weather clouds broken up by strong winds. No precipitation unless winds steady NE to S, then cumulus fractus of bad weather.

E—STRATOCUMULUS at sunset
Low water clouds of dark Indian-red color. Precipitation in 12 to 20 hours if winds from NE to S. Other winds bring overcast sky.

F—ALTOCUMULUS and CIRRUS at sunset
Middle water to ice clouds with brilliant white background. Precipitation likely within 20 to 24 hours if winds are from NE E to S.

G—STRATOCUMULUS at sunset
Low to middle water clouds with gold, pink, amber, lavender or rosy background. No precipitation likely within 20 to 24 hours.

may be called a water cloud, although the upper parts have ice particles (Plate 1–B). Because of their shape, these clouds are often mistaken for cirrocumulus (Plate 2–A). Altocumulus are darker than cirrocumulus, although the shapes are similar.

Stratocumulus are water clouds (Plate 1–C). Their precipitation can be rain or snow. The upper parts contain some ice particles. Often, this formation appears to be lighted from behind. They are seen mostly in waves, rolls, or a series of undulations, sometimes with spaces between the rolls. The most common type is the stratiformis (Plate 1–C), with masses of rolls which are wavy in extended sheets. Often, this is the frontal edge of a cold front.

Nimbostratus clouds are water clouds, ranging in color from light to dark grey (Plate 3–C). They appear in sheets and seem diffuse from continuous precipitation. This formation blots out the sun. Under these clouds may be seen ragged low clouds called "scuds," which sometimes merge with the nimbostratus clouds. While called water clouds, nimbostratus, like most clouds, have ice particles in their upper regions. The precipitation from nimbostratus clouds usually, but not necessarily, is of lengthy duration.

Stratus clouds are light to dark gray, usually uniform in appearance (Plate 2–C). Frequently, they are to be seen in ragged patches. The precipitation from stratus clouds usually is a drizzle, light rain, or snow, and may last for several days. Often the sun can be seen through these clouds. Their light-gray color, however, gives them an appearance different from other clouds, and they cannot show a halo. Stratus at sunset can be gorgeous for dramatic color and sweep.

The principal difference between stratus and its cousins nimbostratus and cumulonimbus clouds is that the latter two produce heavy precipitation, while precipitation from stratus clouds is weak but of longer duration. This is one reason why the two types are often mistaken. Nimbostratus and cumulonimbus clouds are preceded by other formations in the sequence, whereas stratus clouds develop directly into stratus.

Cumulus clouds are the spectacular ones, as described earlier. Depending on their relative position to the sun, the edges stand out brilliantly, or dark. Their bases are mostly flat or horizontal and are darker than the upper portions.

Cumulus are termed water clouds and, like many other types, their tops have ice particles. Their precipitation mostly is in the form of showers of short duration. They can be distinguished from stratocumulus (Plate 1–C) and altocumulus (Plate 1–B) because cumulus always are separated, detached, in towering or dome shapes. Some resemble snowy cauliflowers. Sometimes they seem to merge and are taken to be other formations.

Altocumulus and stratocumulus can develop into cumulus. Cumulus fractus of *bad weather* (Plate 8–D) often are seen with the other precipitation clouds. Cumulus clouds develop mostly from convection, or rising air, cooling as it moves upward. Cumulus fractus of *good weather* (Plate 3–A) are clouds mostly in some dome formation, although sometimes they are flattened. This formation is a fair weather cloud and never gives precipitation.

Cumulonimbus clouds are heavy and dark, towering like huge mountains. Their tops usually are smooth, sometimes shaped like an anvil or vast plumes (Plate 7–B). These are water clouds, their tops having some ice particles, and often beneath them are ragged cloud formations, or "scuds." Cumulonimbus frequently look like cumulus congestus. They prevail mostly in middle and low latitudes, and are rarely seen in the polar regions.

The local-type cumulonimbus cloud storm can develop in a short period, with rain or hail falling over relatively small areas. Local winds can be strong. When preceding a cold front, the precipitation from cumulonimbus is usually heavier and of longer duration, since these formations tend to appear in a series.

Cumulonimbus clouds, unlike most formations, have only one subtitle. Cumulonimbus capillatus (Plate 7–B) often is combined with mammatus (Plate 7–G), and the precipitation is heavy. Such clouds frequently bring dangerous winds and hail and frequently are seen before and after tornadoes. Often, cumulonimbus clouds are mistaken for nimbostratus when a warm front is not far away. But when thunder is heard and lightning is seen, then the clouds are *not* nimbostratus.

Lenticularis clouds have elongated, well-defined shapes similar to almonds or eggs. They form on the lee sides of mountains, and do not foretell any particular changes in the weather.

CLOUD SUBTITLES

Some of the more frequent subtitles encountered include:

Stratocumulus *translucidus*: Descriptive of a variation of a stratocumulus cloud. Broken down, translucidus comes from *lucidus*, bright and shining, and *trans*, throughout. Hence, a stratocumulus cloud is one that is bright and shining throughout.

Opacus: Derived from "opaque," meaning a milky translucent variety that shows a measure of light. Thus, stratocumulus opacus clouds are clouds that are thin, with the light readily seen as if coming from behind the cloud, at times remindful of a milky-white glass.

Perlucidus: Lucidus means almost the same as in *translucidus*. Only in parts of the cloud formation is light seen through or between (passage through) clouds. Thus we have cumulus *per*lucidus.

Fractostratus of bad weather: Derived from "fracture" (*fracto*) and "spread out" (*stratus*). Hence, this term is applied to clouds that are broken up by winds and are uneven and separating. Both "good weather" and "bad weather" clouds can be seen fracturing, or breaking up.

Stratus also is applied when cumulonimbus or other formations begin to spread and cover the sky, as in nimbostratus, stratocumulus, cirrostratus, and altostratus.

A "nimbus" cloud always implies rain. A nimbus cloud is gray and dense.

CLOUDS FORETELL the weather as clearly as a voice announcing an event about to happen. The winds usher in this event. First, identify the cloud, understand it, *know it well*, then determine the direction of the wind.

Keep in mind that weather changes almost always are heralded well in advance and usually come with the wind.

A sure sign of approaching rain is a changing procession of the four basic formations: cirrus, cumulus, stratus, and nimbus.

A comparison of morning and evening skies with the illustrations here will make you into an amateur weather prophet.

There is considerable fun to be had, as well as accuracy in the predictions, by forecasting weather by wind directions:

Fair weather usually comes with northwest, west, and southwest winds. Winds from northeast, east, and south bring unsettled weather (see Plate 9).

If there is rain in the morning with winds from northeast to south, and the wind begins to shift to western points, then the rain will soon cease.

If the sky is cloudy and the wind shifts from southwest to southeast, or from northwest to northeast, then a squall can be expected.

If the sky is clear and the winds begin to shift back and forth between southeast and southwest, then unsettled weather with possible rain is on the way.

If there is an early morning fog, frost, or dew, and the previous night was clear, then there will be no rain for the day.

Fronts are boundary lines of discontinuity set up between moving masses of air of different temperature, humidity, or winds. These contrasting air masses are indicated as "highs" on weather maps. Often, low-pressure circulations, or "lows," will form along the fronts. The power and the vast sweep of fronts on the earth are tremendous, yet these fronts cannot be seen, except in the cloud formations that indicate their approach. They usher in the cold "spells," the heat "waves," the droughts. They move off-normal weather to large areas of the earth, hold it there, sometimes for many

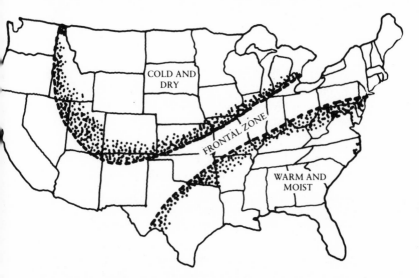

Fig. 4 The boundary line between air masses of different temperature, humidity, or winds is called a front.

days, and move it out. However, specific, well-defined clouds do appear in sequences, telling us that a front is on its way toward us, and what kind of front—whether cold or warm—it is. That the clouds tell us in advance of these far-reaching changes in the weather is another marvel of the communications possible between man and his environment!

A "high" is an immense mountain of air, with winds spiraling clockwise around its center.* Generally, highs, or high-pressure areas, bring fair weather. They form over cool regions of the earth, where the cool and dense air sinks toward the ground—causing the air mass to become "heavy." Thus, the name *high pressure*. Such an air mass may move and cover most of the United States, or any other large land area or ocean.

"Lows" form *between* two areas of high pressure—usually along a frontal zone. Winds spiral counterclockwise* around lows, with light air near the center rising upward from the ground. This rising motion causes lows to bring us the characteristic unsettled weather.

There are four types of fronts: cold, warm, occluded, and stationary.

Cold front: a movement of a cold, or cool, air mass.

Warm front: a movement of a warm air mass.

Occluded front: when a cold front overtakes a warm front and "shoves" it along, the cold usually predominates. An occluded front thus formed is a composite front of cold and warm air. The cold front blocks, or occludes, the effectiveness of the strictly warm front.

Stationary front: an almost motionless boundary between air masses, often remaining in the same position for several days.

COLD FRONTS

For the United States, northern Canada is a vast icebox, or refrigerator, that holds cold air ready to move when the winds begin. As if on a huge, sloping ramp, this cold air travels from

* Wind directions opposite in Southern Hemisphere.

the north down toward the south. It is in the shape of a wedge, with a "sharp" edge—of course invisible, except for the clouds that tell us when the movement is beginning. Cold fronts move rapidly, with greater speed than do warm fronts.

Cold fronts can be preceded by thunderstorms, tornadoes, hail, and sometimes simply by a brisk wind and an overcast sky.

There are continental, or land, cold fronts; and maritime, or ocean, cold fronts. Most of the cold weather in the United States comes from the continental polar fronts, the Canadian icebox. The maritime cold front gives the Pacific Northwest its unusual weather. This area is the rainiest in the United States, and day-to-day weather forecasting there is very difficult. The northeastern part of the United States also is affected by maritime cold fronts.

Usually, the second day after the cold front arrives is colder than the first day—or, in the summertime, cooler.

Watching the entire cloud procession, or sequence (Plate 10) of a cold front in prospect is interesting and enlightening. To

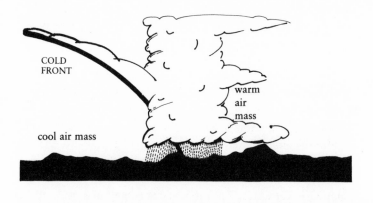

Fig. 5 Cold Front—Drawn with blue triangles on a weather map. It is the leading edge of a cold air mass pushing under a warm air mass, and moves in the direction of the triangles. Weather ahead of the front can be stormy, but usually clears rapidly behind it. Temperatures behind the front will also be colder.

watch these formations advancing evokes a thrill, as if messengers are bringing the news of what is in store or what is not in store for the viewer's locality. It is something like watching the development of a seed into sprouts and, finally, the fruit-bearing plant, complete with colorful blossoms. Unless something *interrupts* the sprouting, the development of stems, leaves, flowers, and the fertilization, we know for certain that the plant will bear fruit because such a progression of development can mean only that.

In predicting the arrival of a cold front, we watch for this order of cloud formations:

First, we may see some cirrus clouds (Plate 10–B). This formation can last for several days or less. It is mainly a "fair weather" cloud, either in advance of a cold front or a warm front.

The next formations in the sequence can vary dramatically, depending upon the speed and movement of the front and the time of year. Cirrocumulus, followed by altocumulus or altostratus (Plate 10–C, D, E), may occur. Stratocumulus (Plate 10–F) would follow—the *real* warning cloud of rain and sometimes gusty winds. The actual rain clouds would be the cumulonimbus (Plate 10–H) or nimbostratus (Plate 10–G). Precipitation may be locally heavy, particularly with cumulonimbus clouds.

The cumulonimbus is the true *thunderhead*, a powderkeg of immense energy which releases its fury with downpours of rain and hail, and spectacular sound and light displays of thunder and lightning. These storm clouds are particularly numerous immediately ahead of a strong and fast-moving cold front. They may organize into well-defined *squall lines*, "strung out" like an enormous string of huge atmospheric "popcorn kernels." One thunderstorm may end, only to be followed by a second or even a third before the front finally rushes through.

In sharp contrast, a weak, slow-moving cold front may usher in only a change in the winds—with merely a few clouds heralding its approach and passage.

The strength of a cold front is determined by how sharply the two air masses on either side of it differ.

A very cold air mass plowing into a very warm one, like a gigantic bulldozer, will make for a strong and powerful cold front. The warm air will be driven upward by the colder air as if being thrown over its shoulder, and deep cumulonimbus clouds may develop. The weather changes with this advancing system will happen quickly, and the front will pass almost as fast as it came, ushering in cooler and less humid air.

All cold fronts, strong or weak, will deliver some sort of weather change in their wake, from a barely noticeable shift in the winds to an icy metamorphosis from tropical warmth.

Depending on the time of year, cold fronts move in cycles that average about five to seven days. This is why in the old days it was claimed that changes in the weather came with the changes of the quarters of the moon. (Many people still think so.) In fact, the moon plays no part in these weather cycles.

WARM FRONTS

Unlike the cold front, a warm front is much gentler in nature, and likes to take its time in its march across the land. This is because the warm air behind the front, in trying to replace the colder air ahead of it, is fighting a struggle comparable to that of David against Goliath. In this case, David, the *warm air*, is destined to win the battle, but must overcome a mighty adversary in Goliath, the *cold air.*

The warm air gradually shoves its way over the colder air, in effect gently "coaxing" the cold air mass to move. In the process, a virtual smorgasboard of cloud types develops ahead of the front. Since a warm front travels much more slowly than a cold front, the notice of its coming is more clearly defined in the various formations of the clouds. The depth of a warm front can range up to 1,000 miles or more, and therefore a careful observer can usually spot the first warning signs of its approach at least one or two full days before the precipitation arrives.

Fig. 6 Warm Front—Drawn with red semicircles on a weather map. It is the leading edge of a warm air mass sliding over a cold air mass, and moves in the direction of the semicircles. Cloudy, wet weather is often found ahead of the warm front, and warmer, partly cloudy weather follows behind it.

The first clouds to be seen in advance of a warm front would be cirrus (Plate 11–A), followed by cirrus spissatus (Plate 11–B) spreading over and covering the sky.

Sometimes, but not always, the next formation to appear will be the cirrocumulus, or mackerel sky (Plate 11–D). Folklore explains this cloud: "Mackerel sky, not yet wet, not yet dry."

Often confused with the cirrocumulus, altocumulus clouds (Plate 11–E) are lower in the sky and have darker, more voluminous rolls. They are "threateners" rather than true precipitation clouds.

Following in the procession will be the altostratus clouds (Plate 11–F), when the sun may appear as behind frosted glass, or as if it were a vast cocoon glowing from inner light. Then at sunset we are likely to see a dark Indian-red sky (Plate 11–G), or a yellow sky, or the brilliant white sunset (Plate 11–H). The background of these formations may be altocumulus, stratocumulus, or stratus—and they are what we call color skies, the ones that proclaim the closeness of wet weather.

The actual rain- or snow-producing clouds—the nimbostratus (Plate 11–I, L) and possibly the cumulonimbus (Plate 11–J) — are the next clouds in the sequence heralding the warm front's approach. Precipitation from these clouds may be of long duration and occasionally heavy. Stratus clouds (Plate 11–K) may bring light drizzle or, in winter, light snowflakes, lasting for several days.

Once the warm front passes, skies will begin to clear and the temperature will rise. It is here that the intimate link between warm and cold fronts is revealed, for a warm frontal passage is almost always followed by the approach of a cold front. The success may not be immediate, and in fact often takes several days, depending on the time of year and the strength of the weather systems involved.

STATIONARY FRONTS

A stationary front is not a warm front or a cold front. It is an almost motionless collision of air masses, with neither side strong enough to win the battle for supremacy. It is as if the stormy troposphere were holding a quiet, but uncertain, conference over some tormenting of the earth.

Rain, drizzle, fog, snow, clouds, sunshine—you name it, almost any kind of weather can be found along a stationary front, although clouds and precipitation are usually the most popular guests at this party. As long as the front doesn't move, there will be very little change in the existing weather, whatever form it may take. These fronts are especially prevalent during the summer months.

Eventually, stationary fronts either dissipate, as the air masses on either side acquire similar characteristics, or evolve into warm or cold fronts.

If the warmer air mass finally gains the upper hand, a warm front is formed, and you can look for light precipitation followed by warmer temperatures.

If the colder air mass prevails, the resulting cold front will usher in a change to cooler temperatures.

OCCLUDED FRONTS

Sometimes a cold front will closely follow a warm front, as if "stalking" it. When the cold front, which is wedge-shaped, overtakes the warm front, the warm front is shunted aloft. This shunting cuts off the progress of the warm air. The cold front edge and the warm air continue to move along, together, the cold pushing the warm. This moving line of contact is the occlusion.

The duration of an occluded front—the cold edge and the warm edge moving along together—is uncertain, as is the case with the stationary front. The presence of an occluded front is "told" mainly by the various stratus clouds. These fronts usually bring precipitation, but, as a rule, it is not heavy.

6 THUNDERSTORMS AND LIGHTNING

THE THUNDERSTORM is one of nature's most awesome creations. It strikes fear in the hearts of many, and wonder in the minds of the curious.

Its vast sweep of power and energy is nothing short of electrifying. Indeed, a single flash of lightning from its mighty clouds packs enough power to light an entire town. Even our most efficient electric power systems would be hard pressed to match the strength and capacity of that single bolt from the sky. Yet we cannot harness this atmospheric energy; we can only speculate as to its hows and whys.

The cumulonimbus cloud is the lifeblood of a thunderstorm. Its majestic, towering form can reach heights of 40,000 feet or more. This remarkable vertical development is the result of strong uplifting of the air, which can come about in several ways. Local heating can induce rising air motion by creating warm air parcels near the earth's surface, which will rise upward like balloons. Or a cold front can force air upward by plowing into the air mass ahead of it. Sometimes, all it may take is a temperature difference between land and water to induce rising motion, particularly in coastal regions.

Once the air rises, it cools, and moisture in it may condense. Small, puffy cumulus clouds (Plate 12–A) may be the first clouds to form. If the atmosphere is *unstable,* then the air within the cumulus will continue to rise, forcing even more condensation of water vapor, and subsequently a more deeply developed cloud.

This instability of the atmosphere simply means that the rising air, although cooling in its upward journey, remains warmer than the air surrounding the cloud. Thus it continues to rise. This is how the sometimes violent, chimneylike updrafts develop within a cumulonimbus.

Since the upper reaches of these massive clouds are actually above the freezing level, ice crystals are the dominant cloud particles high up, while liquid water droplets comprise the lower territory. The strong updrafts of the developing cloud invoke equally strong downdrafts in the mature cumulonimbus at the onset of precipitation.

Speaking of precipitation, there's usually no shortage of it with a full-blown thunderstorm. The powerful up-and-down motion within the cumulonimbus smashes and blends ice particles and water droplets together until they are heavy enough to fall to the ground below. Hail, a unique product of the cumulonimbus, is often mixed with the rain as it reaches the earth.

THUNDER AND LIGHTNING

Watching a thunderstorm in action is almost like the 4th of July revisited. Our senses are touched by nature's own fireworks.

Lightning *causes* thunder. But what causes lightning?

Electrically charged ice crystals and water droplets within the cumulonimbus are in constant, violent collision. In the same way that our bodies build up static electricity through friction with a carpet, the cloud develops into a gigantic "battery," by experiencing a *separation* of electrical charge; positive charges accumulate near the top of the cloud, and negative charges near the bottom.

Eventually, the cloud must release some of this electrical energy in the form of an electrical discharge, called lightning. It can happen within the same cloud, between two different clouds, or between the cloud and the ground.

The flickering of lightning reveals its true nature, having many components rather than a singular flash. Electricity is "drained" back and forth through the lightning channel several times within only a fraction of a second.

There are several types of lightning: sheet, streak, forked, beaded, ribbon, heat, and even "ball" lightning. Sheet lightning appears as a flash of glowing white light within the cloud.

Streak and forked lightning are both cloud-to-ground flashes, which appear to be perfectly straight and branched, respectively.

Very strong horizontal winds can cause streak lightning to appear as several parallel luminous bands, or ribbon lightning.

Ball lightning is rare, and barely understood. It appears as a red luminous ball floating in midair.

Heat lightning is the flashing seen in a thunderstorm at a distance too far away for its crash of thunder to be heard.

Thunder is caused by the intense heat surrounding the lightning channel. Sound waves are created as the air close to the lightning bolt is instantly heated and forced to move very fast, thus producing a "sonic boom" of sorts.

It is interesting to judge the timing factors in a thunderstorm, based solely on an observation of the thunder and lightning. Light travels at the rate of 186,300 miles per second, while sound travels at approximately 1,100 feet per second. So the glaringly brilliant light produced in lightning reaches us almost simultaneously with its arcing, while the thunderous sound comes more slowly. When storms are relatively close by, we can determine their distance from us by counting slowly from the moment of flashing until we hear the thunder.

If, between the flashing of the lightning and the sound of the thunder, we have counted to five seconds, then the lightning flash was one mile away.

THUNDERSTORM CLOUD SEQUENCES

Cloud sequences in advance of a thunderstorm will vary, depending on the nature of the storm's origin (see Plate 13).

The *local* thunderstorm, developing on a hot and humid

summer afternoon, will start as a billowing white cumulus cloud (Plate 12–A) in the late morning or early afternoon. It will begin to darken and build up as the day progresses, gradually evolving into the towering cumulonimbus (Plate 12–E).

Soon, a few peals of thunder are heard, along with some threatening growls, the first stirring of the monster in the skies. The storm is developing. Usually this type of thunderstorm has an average life of about thirty minutes to an hour. There may be some gusty winds and brief heavy rains with hail.

The great thunderstorms, the ones which often cause extensive damage, are produced in advance of a cold front, possibly in squall lines angrily racing in long strides across the earth.

By watching the cloud formations, we can more accurately foretell the coming of this type of thunderstorm. The air gets colder as it advances closer, the result of strong downdrafts within the outer walls of the cumulonimbus. The winds begin to grow in intensity, often in gusts up to 50 miles an hour (see Figure 12, Beaufort Wind Scale for Land).

A cold front thunderstorm may last for hours, with heavy rain, dangerous lightning, and terrific winds. Once the front passes, though, the storminess should quickly come to an end.

WHAT TO DO ABOUT THUNDERSTORMS

Many summer days are thunderstorm days. The summer thunderstorms that come lumbering in on hot, muggy afternoons and evenings to soak heat-parched streets and lawns, the fields and forests, are welcome arrivals in June, July, and August—unless we are caught in one. They can be very dangerous.

Estimates are that several hundred people are killed each year by lightning in the United States. Many times that number survive injuries received from bolts. Holes four inches in diameter have been found in lightning-blasted metal.

The newspaper on the following morning may report that pleasure boats were capsized and their occupants drowned, golfers had been struck and killed, and that other tragedies

PLATE 9

Cirrus *(Good Weather)*

This formation of ice clouds foretells good weather if winds are from southwest to north. But steady wind from northeast to south causes clouds to thicken, cover the sky, and warn of advancing warm front and precipitation.

Cirrocumulus *(Uncertain Weather)*

This formation without winds precedes an overcast sky. If followed by winds from west to north no precipitation is likely, but with winds from northeast to south precipitation may be expected within the next 24 hours.

PLATE 10

Cloud Sequence Marking Approach of a Cold Front

A.—STRATOCUMULUS opacus

Immediate threatener of bad weather from a sprinkle to heavy precipitation. If at head of cold front, gusty winds or thunderstorms.

B.—CIRRUS fibratus

Ice clouds. Good weather if winds from W NW to N. Precipitation likely within 20 to 30 hours if winds steady from NE E to S.

C.—CIRROCUMULUS

Ice clouds. Precipitation likely within 15 to 20 hours if wind NE to S. Early summer A.M. often afternoon thundershowers. Other winds overcast.

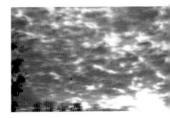

D.—ALTOCUMULUS perlucidus

Water and ice clouds. Some precipitation likely within 15 to 20 hours if wind steady from NE to S. Other winds bring overcast sky.

PLATE 10

E.—ALTOSTRATUS translucidus

Precipitation likely within 10 to 20 hours if wind is steady from NE to S. Sun appears behind frosted glass. Other winds overcast sky.

F.—STRATOCUMULUS stratiformis

Immediate threatener of bad weather from a sprinkle to heavy precipitation. If at head of cold front, gusty winds or thundershower.

G.—NIMBOSTRATUS

Rain or snow clouds. Precipitation of long duration if winds NE to S, or shorter duration if winds are from SW W to N.

H.—CUMULONIMBUS

Precipitation likely and soon coming usually from SW W to N. Distant clouds often show an anvil-shaped cirroform cap.

I.—CUMULONIMBUS mamma

Seldom seen low to middle clouds. Associated with severe wind squalls, hail, heavy precipitation, tornadoes, and thunderstorms.

PLATE 11

Cloud Sequence Marking Approach of a Warm Front

A.—CIRRUS fibratus

Ice clouds. Good weather if winds from W NW to N. Precipitation likely within 20 to 30 hours if winds steady from NE E to S.

B.—CIRRUS spissatus

Ice clouds. Good weather if winds from W NW to N. Precipitation likely within 20 to 30 hours if winds steady from NE E to S.

C.—CIRROSTRATUS covering sky

Ice clouds cover sky. Precipitation likely within 15 to 25 hours if winds steady from NE E to S. Other winds bring overcast sky.

D.—CIRROCUMULUS

Ice clouds. Precipitation likely within 15 to 20 hours if wind NE to S. Early summer A.M. often afternoon thundershowers. Other winds overcast sky.

E.—ALTOCUMULUS perlucidus

Water and ice clouds. Some precipitation likely within 15 to 20 hours if wind steady from NE to S. Other winds bring overcast sky.

F.—ALTOSTRATUS translucidus

Precipitation likely in 10 to 15 hours if winds steady NE to S. Sun appears to be behind frosted glass. Other winds bring overcast sky.

PLATE 11

G.—STRATOCUMULUS at sunset
Low water clouds of dark Indian-red color. Precipitation in 12 to 20 hours if winds from NE to S. Other winds bring overcast sky.

H.—ALTOCUMULUS and CIRRUS at sunset
Middle water to ice clouds with brilliant white background. Precipitation likely within 20 to 24 hours if winds are from NE E to S.

I.—NIMBOSTRATUS
Rain or snow clouds. Precipitation of long duration if winds NE to S, or shorter duration if winds are from SW W to N.

J.—CUMULONIMBUS capillatus
Precipitation likely and soon coming usually from SW W to N. Distant clouds often show an anvil-shaped cirroform cap.

K.—STRATUS (Note tower fading in cloud)
Winds from NE to S may bring heavy precipitation. Other winds bring only light drizzle or an overcast sky.

L.—NIMBOSTRATUS
Rain or snow clouds. Precipitation of long duration if winds NE to S, or shorter duration if winds are from SW W to N.

PLATE 12

Cloud Sequence of Developing Thunderstorm

A.—CUMULUS humilis

Fair weather clouds if they show no vertical development. Can build up and develop into cumulus congestus or cumulonimbus clouds.

B.—ALTOSTRATUS translucidus

Precipitation likely in 10 to 15 hours if winds steady NE to S. Sun appears to be behind frosted glass. Other winds bring overcast sky.

C.—ALTOCUMULUS translucidus

Water and ice clouds. Some precipitation likely within 15 to 20 hours if wind is steady NE to S. Other winds bring overcast sky.

D.—CUMULUS congestus

If clouds form from SW to NW precipitation with gusty winds and thunderstorms or only wind squalls are likely within 5 to 10 hours.

PLATE 12

E.—CUMULONIMBUS

Precipitation likely and soon coming usually from SW W to N. Distant clouds often show an anvil-shaped cirroform cap.

F.—STRATOCUMULUS opacus

Immediate threatener of bad weather from a sprinkle to heavy precipitation. If at head of cold front, gusty winds or thunderstorms.

G.—CUMULONIMBUS mamma

Seldom seen low to middle clouds. Associated with severe wind squalls, hail, heavy precipitation, tornadoes, and thunderstorms.

H.—CUMULONIMBUS

Precipitation likely and soon coming usually from SW W to N. Distant clouds often show an anvil-shaped cirroform cap.

PLATE 13

Stratocumulus *(Weather Threatener)*

With winds from west to north, this formation threatens unsettled weather, preceding a cold front. But winds from northeast to south may change the clouds from stratocumulus to nimbostratus, thus making precipitation likely.

Cumulus Congestus *(Precipitation Likely)*

These low water clouds with winds blowing from the northeast to south indicate precipitation within twenty-four hours. But with winds from west to north you can expect an overcast sky and possibly a local thunderstorm.

occurred. A thunderstorm, with its sudden advent of gale-force winds and lethal charges of lightning, is a formidable visitor.

Most of the fatalities due to thunderstorms could be avoided. There is actually nothing "sudden" about a "sudden storm," and if people will only understand what storms are and how they can be avoided, the damage toll would be greatly reduced.

The best protection against being caught in an exposed place during a thunderstorm is to be able to recognize the cloud formations preceding it. Remember to keep a wary eye on cumulus clouds that darken and thicken rapidly, especially during the afternoon.

Here are some good lightning safety rules:

If at all possible, stay inside your house during a thunderstorm. Don't bathe or shower until the storm is over.

If you are driving a car during a thunderstorm, stay in it. The metal exterior will act as a protective shell against lightning hitting *inside* the car.

If a violent storm catches you out of reach of a building or automobile, then lie prone on the ground near an embankment or ditch, not in the shelter of a tree.

Swimmers should *get out of the water* if a storm approaches. It is one of the most dangerous places to be.

Boaters should watch for the build-up of threatening clouds, and head for safe anchorage if a storm looks likely. Both lightning and high winds can pose a serious safety hazard.

7   HURRICANES AND OTHER STORMS

THE MOST devastating storm that visits our earth is the hurricane, wild child of the tropics, wandering with its terrible violence deep into the temperate areas of our land and seas.

Wild as it is, the hurricane is the most orderly in its form and behavior, when we compare its activity with all the other vagaries and patterns of weather. The hurricane's orderliness, its "shape," and duration, are as precise as a perfectly geared, well-oiled, expertly maneuvered machine.

If a hurricane were wholly visible as a solid, and we could slice it in half from top to bottom, we would have two almost perfectly symmetrical parts.

The hurricane winds spin as fast as 200 miles an hour, yet its center is a nearly perfect calm! This center is its "eye," surrounded by nimbostratus clouds—like a gargantuan doughnut. When we find ourselves directly under this "eye" we can see a disc of clear blue sky above as we stand in an almost breezeless area, the raging winds encircling this quiet place. This situation is one of the weirdest that can be found in weather behavior.

The "eye" of a hurricane is as remarkable in its curious behavior as is the storm that roars around it, the "eye" having an average diameter of 15 miles. The hurricane, when fully grown, reaches a diameter seldom less than 50 miles and can expand to a diameter as great as 150 miles or more. The diameter of the area affected by such a storm may be 500 miles.

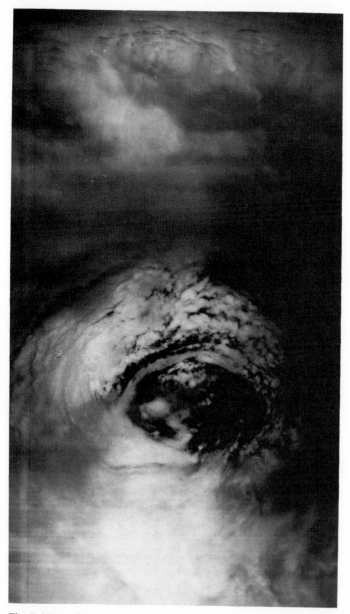

Fig. 7 View of Hurricane Betsy, 90 miles north of Grand Turk Island in the Caribbean, taken from 11 miles above the earth by a United States Air Force reconnaissance aircraft September 2, 1965. Note the classical circular structure and central eye of the hurricane. *Photo courtesy National Oceanic and Atmospheric Adminstration.*

Originally, the term hurricane applied only to tropical storms in the West Indies (the name is derived from the West Indian word *huracan*), but now is applied to *any wind force* exceeding 74 miles per hour. (See Figure 12, Beaufort Wind Scale.)

In the Western Pacific regions this same type of storm is known as a typhoon. It is called a willy-willy near Australia. Hurricanes and typhoons are not to be confused with other tropical weather disturbances which bring winds of high force and considerable rainfall. The hurricane is distinct unto itself.

All hurricanes are born over water. Their life cycle is about ten days. They travel rather slowly, at about 15 miles per hour, and the path they will take can be quite erratic and hard to predict. Satellite pictures help meteorologists forecast their movement, and tell them when to issue warnings for the approach of a hurricane. In the sky you will see a progression of cloud formations in advance of the storm's approach, usually starting with thickening cirrus clouds.

Fig. 8 Coastal areas of the Eastern and Southeastern United States are most susceptible to the hurricane threat.

Once a hurricane reaches a land area, such as the coast of the United States, it begins to die. This is because the great masses of water vapor on which the whirling monster thrives are sharply reduced by the land body, and also because the land offers friction, as opposed to the smoothness of the water surface on which the aerial whirligig seems to have its footing. An island in its path also reduces a hurricane's fury.

All hurricanes originate in lower latitudes, where warm ocean waters can "feed" the growing storms sufficient energy. Their track of movement eventually turns northward (if born north of the equator), and southward (if born south of the equator).

The exact track that the hurricane takes will determine which land areas will be in its path of destruction. Many of these hurricanes, in fact, never reach land areas, and die over the colder waters of the North Atlantic and Pacific. The hurricanes that do hit land areas will deal their most destructive blows to islands and coastal regions.

Fig. 9 Satellite weather photo of tropical storm off North Carolina and the Virginia capes. *Photo courtesy National Oceanic and Atmospheric Administration.*

The accompanying rainfall can be tremendous; 12 to 15 inches in one day on a single locality is not unusual. In the tropics, and on several storm occasions on the southeastern seaboard of the United States, rainfall during the terrible dance of a hurricane has reached more than 20 inches in a single day.

Along with the considerable lashing of the winds and deluges of rain in what sometimes seems to be a solid body of water descending come the raging waves of the sea, which rise quickly and frequently cause enormous property damage and loss of life. Smaller boats and ships at sea can become total victims in the path of a hurricane, and the smaller islands sometimes are left desolations of wreckage, overwhelming floods, monumental landslides, uprooted trees.

When the moon is full, the tides drawn by a hurricane are even more forceful against the shores.

In the Northern Hemisphere, "hurricane season" lasts from June to November, with August and September being the *prime* months. In the Southern Hemisphere, hurricanes become most prevalent during the months of January through April.

The layman can do little about forecasting a hurricane or tropical storm, and must depend on the forecasts of trained meteorologists to prepare adequately for a storm's arrival. By observing the changing progression of clouds in advance of the storm, however, *you* will be able to *see* the storm's approach, and know of its arrival.

Now let us see how the clouds can tell the progress of a hurricane or tropical storm as it comes toward us.

HURRICANE CLOUD WARNINGS

The sequence of cloud formations that precedes a hurricane is similar to the movements of clouds that foretell the approach of a warm front. Remember—*warm fronts, by comparison with cold fronts, move slowly*. Consequently, the hurricane moves with an agonizing slowness, keeping millions of people in suspense. The anxiety of awaiting possible arrival of a hurricane is especially frightful inasmuch as, often, the hurricane will "change its mind," and suddenly veer off to the left or right.

High cirrus clouds will often show up several hundred miles in advance of a tropical storm or hurricane. If the storm continues to move closer, the cirrus will thicken to become cirrostratus (Plate 14−A), and eventually the darker altostratus (Plate 14−B, C, D). When this happens, the clouds are telling us that the storm may be only a few hundred miles away.

If the altostratus further develop into stratocumulus or nimbostratus (Plate 14−F, G) in advance of a tropical storm or hurricane, then the storm is very close and rains will begin to fall and the winds will pick up in intensity. Cumulonimbus (Plate 14−H) clouds imbedded in the storm may bring thunder and lightning, and there may even be tornadoes.

The duration and intensity of storms such as these can vary greatly, and although they will lose their strength rapidly once landfall is made, the subsequent clouds and rainfall may hang on for days.

Remember, hurricane forecasting is best left to the professionals. Listen to their forecasts and warnings before making your own sky analysis.

TORNADOES

People often confuse tornadoes with hurricanes, or vice versa. Actually, the difference between the two is tremendous. Unlike the hurricane, a tornado is rather small in size and short in duration. In fact, tornadoes can, and often do, originate *within* a hurricane, but a hurricane can never arise from a tornado.

Tornadoes are closely related to thunderstorms, inasmuch as atmospheric conditions that produce a severe thunderstorm can also breed a tornado.

When masses of cold, dry air collide with masses of warm, moist air, conditions often become ripe for severe weather. The atmosphere can be very *unstable* in this collision zone, leading to strong vertical air motion. This rising motion is the

Fig. 10 Tornado, photographed near Rainview, Texas, May 27, 1978. *Courtesy National Oceanic and Atmospheric Administration.*

lifeblood of a thunderstorm, and ultimately of a tornado.

The topography east of the Rockies to the mountain ranges near the East Coast is more favorable for tornadoes than is any other part of the United States. This is because there are no barriers to the movement of the cold and warm air masses as they slide and converge from the north and the south. This *collision* of air is most pronounced during the spring, when both the polar and tropical air masses are "battling it out" for dominance. Thus, tornado frequency reaches an average peak in April, May, and June.

This is not to say that tornadoes don't occur at other times of the year, and in other places. In fact, most parts of the country have, at one time or another, experienced tornadoes,

Fig. 11 United States tornado frequency is the highest in the Great Plains, especially during spring and early summer, but tornadoes can—and do—occur in other parts of the country, and frequently at different times of the year.

and they have occurred in November as well as in August, or most any other month of the year.

The tornado—also called a "twister"—is a funnel-shaped cloud extension that reaches the ground. It "drops out" of a cumulonimbus cloud, usually during a severe thunderstorm. Sometimes indicative of tornadic conditions, and often seen afterwards, is cumulonimbus mamma (Plate 7–G). This cloud appears as many drooping, heavy sacks, sometimes very pronounced and darkly angry. The actual visible beginning of a tornado is a lowering, spoutlike, seemingly elastic cloud—a funnel cloud, an extension from the mother cloud of its birth.

This tentacle dips toward the earth to hover, like an elephant's swaying trunk, 50 or 100 feet above the ground. When it touches the earth the devastation is sometimes incredible. In addition to its whirling wind speed of 50, 100, or even 200 miles per hour, the extreme centrifugal force and very low barometric pressure can cause almost anything to happen. Buildings can literally explode. Houses have been lifted in their entirety and moved many feet; wheat straws have been driven into tree trunks; trains lifted from their rails; structures and vegetation ripped apart.

Also, curious things have happened during a tornado. Persons have been lifted many feet, transported great distances and lowered gently to the earth, none the worse for the experience. Poultry have been found, after a tornado, apparently unharmed, but plucked almost entirely of their feathers!

Along the ground, tornadoes travel in paths averaging 300 to 400 yards in width; some have been as broad as a mile, or more. The path of a tornado is relatively short, from just a few miles to 50 miles. In rare instances, they have traveled even further before either dissipating or lifting from the ground. Their speed of travel is from 10 to 50 miles per hour.

On a small and quite harmless scale we see the principle of a tornado in spring and summer. Suddenly, out of "nowhere" comes a vagrant air current that draws up dust and papers, spinning counterclockwise. These tiny "dust devils" dance away and quickly disappear. Actually, they are miniature tornadoes.

Although the air conditions known to breed tornadoes can be defined with some certainty, tornadoes themselves are extremely difficult to predict, even with the use of sophisticated radar and satellite technology. Meteorologists routinely issue watches and warnings for the likelihood of tornadoes in specific areas, and by doing so give us the best defense possible against these small storms of such awesome destruction.

From the soothing zephyr to the raging cyclone, all winds are dramatic. The various winds do not simply "puff" out of nowhere, but arise from differences in temperature, pressure, and terrain all over the world.

UNUSUAL WINDS

If a wind can be termed "picturesque," certainly the one called "Chinook" in the Rocky Mountains and "Foehn" in the Alps deserves the label. This wind is a feature of the eastern slopes of mountains. Air moves up the western slopes, spills over and descends the eastern slopes, compressing and thus warming at the rate of about 5.5 degrees F per 1,000 feet as it pours downward. Such winds, which are very dry, have been known to raise the temperature as much as 30 degrees F within fifteen minutes! The Chinook, in winter, is remarkable for its rapidly moderating effects, making snow vanish with surprising speed.

The Mistral, a northerly wind capable of great violence when the atmospheric pressure differences increase, is cold and dry, confined to the northwest Mediterranean, particularly in the Rhone delta. The Mistral is marked by intense cold and dryness, clear skies and brilliant sunshine. The Mistral has been recorded as blowing as many as 175 days within a year.

The Sirocco, also an habitué of the Mediterranean, is a hot, dry wind of Sicily and south Italy, while the Leveche, hot and dry, favors the Spanish coast and about ten miles inward from Valencia to Malaga.

Madeira experiences the Leste, hot and dry dust-bearing (much like the Leveche), which is generally a winter wind.

The "fifty wind," from the Arabic *khamsin*, meaning "fifty," is a late spring visitation upon Egypt. It is a southerly, hot, dust-laden force that usually lasts 50 days.

Monsoon is another important wind having, to the western world, exotic connotations. The name derives from the Arabic *mausim* or *mawsim*, meaning season. The Monsoon blows approximately six months from northeast and six months from southwest over the Arabian Sea, but actually is much more extensive. We hear of it more as it influences southeast Asia. For instance, the Monsoon is a vital factor in the livelihood of India's people. A "bad" Monsoon can produce famine, a "good" Monsoon promises abundance, the one bringing little, if any, rainfall, the other bringing sufficient rainfall. The causes and total extent of the Monsoon circulation (in many parts of the world) are too complicated for discussion here.

HOW TO ESTIMATE VELOCITY OF WINDS ON LAND (BEAUFORT WIND SCALE)

Beaufort Number	Miles Per Hour	Description Winds	Effects Shown
0	0	Calm	Smoke rises straight up.
1	1-3	Light air	Smoke drifts; will not move weather vanes.
2	4-7	Light breeze	Leaves rustle, some vanes move.
3	8-12	Breezes	Leaves and twigs move, small flags wave, vanes move.
4	13-16	Moderate breeze	Branches move, loose paper and dust blows around.
5	17-22	Fresh breeze	Small trees with leaves move, palms bend some.
6	25-28	Strong winds	Large branches move, whistle noise in window cracks and through wires.
7	28-34	Light gales, small craft, flags and lantern signs	Whole trees sway.
8	34-40	Gale	Twigs break, body feels strength of winds.
9	40-49	Gale	Signs blown down, shingles loosen, tree branches break loose.
10	50-54	Gale	Small trees uprooted, loosens shingles, side walls of house may open frame.
11	54-73	Whole gale	Extensive damage to property, trees, etc.
12	73-up	Hurricane	Extensive damage to structure and trees.

Fig. 12 *Prepared by U.S. Department of Commerce, Environmental Science Services Administration*

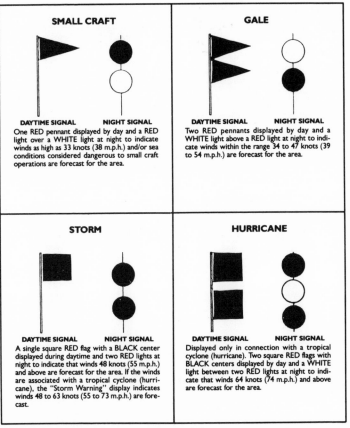

SMALL CRAFT

DAYTIME SIGNAL   NIGHT SIGNAL

One RED pennant displayed by day and a RED light over a WHITE light at night to indicate winds as high as 33 knots (38 m.p.h.) and/or sea conditions considered dangerous to small craft operations are forecast for the area.

GALE

DAYTIME SIGNAL   NIGHT SIGNAL

Two RED pennants displayed by day and a WHITE light above a RED light at night to indicate winds within the range 34 to 47 knots (39 to 54 m.p.h.) are forecast for the area.

STORM

DAYTIME SIGNAL   NIGHT SIGNAL

A single square RED flag with a BLACK center displayed during daytime and two RED lights at night to indicate that winds 48 knots (55 m.p.h.) and above are forecast for the area. If the winds are associated with a tropical cyclone (hurricane), the "Storm Warning" display indicates winds 48 to 63 knots (55 to 73 m.p.h.) are forecast.

HURRICANE

DAYTIME SIGNAL   NIGHT SIGNAL

Displayed only in connection with a tropical cyclone (hurricane). Two square RED flags with BLACK centers displayed by day and a WHITE light between two RED lights at night to indicate that winds 64 knots (74 m.p.h.) and above are forecast for the area.

Fig. 13 Weather Bureau Warning Signals

8 OUR HEAVENLY HUES—
and What They Mean

HART CRANE:

> ". . . the nimble blue plateaus,
> The mounted, yielding cities of the air!"

SHAKESPEARE:

> "Look you, this brave o'erhanging firmament,
> This majestical roof fretted with golden fire."

ROBERT HERRICK:

> "See how Aurora throws her fair
> Fresh-quilted colors through the air."

ON ONE DAY we may see delicate trailing wisps like a horse's silvery tail sweeping against the vivid blue. On another day there might be towering snowy thunderheads, or the canopied dappling of a mackerel sky. Sometimes the clouds are deep gray and forbidding, but not often. We see the clouds as raging flames, or like carvings from rich brass, and we see the fire of diamonds, the deep glow of rubies, the cool light from the heart of an emerald. We find every color, every tone, every hue, and always in harmony, in the clouds.

The many colors of the sky, as we have said, are created by the sun's rays passing through the atmospheric dust and moisture, just as colors are seen when white light shines through a glass prism on a table lamp. Blue is the shortest band in the spectrum and is the predominating color of the sky. Red and pink, the widest bands of the spectrum, are seen most often at sunrise and sunset. This is because the sun's rays must pass through a large horizontal layer of our atmosphere as the sun gets closer to the horizon.

When at morning the sky is gray the hue lasts only a short while. This indicates the end of the passage of clouds (Plate 16–A). This is a sign of good weather for the day. "... morning gray sends a traveler on his way."

The brilliant rose, pink, coral, amber, and lavender show more on the high ice clouds. These clouds, too, are good weather clouds, and the colors indicate good weather for 15 to 25 hours at least (Plate 16–B, C, D).

The dark red and the yellow sky near the end of the sunset apply to lower water clouds mostly. These two colors indicate unsettled weather within the next 24 hours. This also applies to the brilliant white sunset (Plate 16–F), which is an unsettled weather formation.

The gold and amber colors predominate in the middle ice and water clouds. But these colors also appear on the lower clouds and are a sign of good weather.

A green sky at sunset, seldom seen, is almost a sure sign of unsettled weather within the next 15 to 20 hours. This sky is seen more often at sea than on land.

On a hot summer day, when the sun is fiery red as it sets or rises, and there are no clouds, the day, or the one to come, will continue to be hot and the weather fair.

9 CURIOUS UNTRUTHS AND ODD TRUTHS

F OR SO LONG, the moon has been so fixed as a part of
folklore beliefs in weather behavior that to be convincing
otherwise is difficult and may forever remain so. Many people
are adamant in their convictions that the moon positively dic-
tates the proper planting time, proper time to harvest crops,
to "put up" (can or pickle or preserve) vegetables, fruits, and
berries, and to butcher animals for curing the meat.

But it is the *sun*, and not the moon, which is the dominating
force in our weather. The belief that the moon does exert an
effect on the weather at regular intervals doubtless arises from
the moon's quartering changes which occur on the average of
about every eight days. It is pure coincidence that our weather
patterns cycle in more or less the same time frame.

"We are in for rain, because the horns of the moon are
pointed down," we hear people say, sincerely believing this
to be true. Such a moon is supposed to "spill water," like an
upside-down bucket.

"We are in for a dry spell, because the horns of the moon
are pointing upwards and the moon will hold back the water,"
becomes a consequent natural belief.

However, any weather, dry or wet, that may occur on the heels
of such observations is purely coincidental.

But the moon *can* be an aid in foretelling weather from day
to day. The moon's appearance behind certain formations of
clouds can guide us. When the outline of the moon is clean
and sharp, and the stars are clear and twinkling, the weather

PLATE 14

Clouds Sometimes Seen When Hurricanes Are Developing

A.—CIRROSTRATUS covering sky
Precipitation likely within 15 to 25 hours if winds steady from NE to S, or sooner if winds SE to S. Other winds bring overcast sky.

B.—ALTOSTRATUS translucidus
Precipitation likely in 10 to 15 hours if winds steady NE to S. Sun appears to be behind frosted glass. Other winds bring overcast sky.

C.—ALTOSTRATUS translucidus
Precipitation likely within 10 to 20 hours if wind is steady from NE to S. Sun appears to be behind frosted glass. Other winds overcast sky.

D.—ALTOSTRATUS translucidus
Precipitation likely in 10 to 15 hours if winds steady NE to S. Sun appears to be behind frosted glass. Other winds bring overcast sky.

E.—CIRROSTRATUS with halo
Precipitation likely within 15 to 24 hours if wind steady from NE to S. Prismatic effect of sun or moon through ice crystals causes halo.

F.—STRATOCUMULUS stratiformis
Immediate threatener of bad weather from a sprinkle to heavy precipitation. If at head of cold front, gusty winds or thundershower.

G.—NIMBOSTRATUS
Rain or snow clouds. Precipitation of long duration if winds NE to S, or shorter duration if winds are from W W to N.

H.—CUMULONIMBUS
Precipitation likely and soon coming usually from SW W to N. Distant clouds often show an anvil-shaped cirroform cap.

PLATE 15

Cloud Sequence that Often Signals Change from Fair to Stormy Weather

1. FAIR WEATHER SKY.
White fluffy cumulus clouds bring a day of good weather. Morning frost or ground dew is an almost certain sign of no rain for the day.

2. WEATHER CHANGE is foretold by white fleecy cirrus clouds. N to NE winds bring an overcast sky but no rain for 48 hours. SE to SW winds, rain in 24 to 36 hours.

3. UNCERTAIN SKY. This cirrocumulus sky with wind from NE to SW may bring a short late afternoon rain. Other winds bring no rain. This sky favors good weather.

4A. RAIN WARNING SKY. A sky of dark altocumulus clouds against a background of glaring white sunlight at sunset is usually the forerunner of rain in 24 hours.

4B. RAIN WARNING SKY. A dull Indian-red sky at sunset, or at sunrise, warns of rain within the following 24 hours, possibly accompanied by strong winds.

PLATE 15

4C. RAIN WARNING SKY. A morning sky of Indian-red altocumulus clouds usually brings rain, often accompanied by winds, and summer thunderstorms.

5. RAIN WARNING SKY. If the sun seems to be within a halo, behind frosted glass of cirrostratus clouds, or in a cocoon of altostratus, expect rain within 24 hours.

6. RAIN WARNING SKY. Small dark altostratus clouds like these usually bring rain by nightfall. If wind is E to SW then rain is fairly certain within 24 hours.

7A. RAIN AND SQUALL. This sky of cumulus congestus clouds with E and Southerly winds brings showers in summer. Wind shift from W to N brings clear weather.

7B. RAIN OR SNOW. Light gray to black cumulus congestus clouds with E to Southerly winds indicates rain or snow. N to NE winds— light rain. W to N—clear.

8. THUNDERSTORMS. Cumulonimbus clouds are a sure sign of thunderstorms and showers, but a wind shift to W or NW will end rain and bring cooler weather.

A. EARLY MORNING GOOD WEATHER SKY.
A gray sky is the forerunner of fair weather. An early morning fog is always followed by a rainless day if the preceding night was clear.

B. GOOD WEATHER SKY. A faint lavender sky, with high blue above the clouds in early morning or late afternoon, foretells good weather. Seldom seen in winter.

C. CLEAR WEATHER SKY. Whether the weather at the moment be clear or cloudy, this rosy sky at sunrise or sunset will bring good weather the following day.

D. WIND SKY. A golden amber sky foretells of wind rather than rain. A pale yellow sky warns of rain within the following 12 to 24 hours.

E. RAIN WARNING SKY. A morning sky of Indian-red altocumulus clouds usually brings rain, often accompanied by winds, and summer thunderstorms.

F. RAIN WARNING SKY. A sky of dark altocumulus clouds against a background of glaring white sunlight at sunset is usually the forerunner of rain in 24 hours.

G. RAIN WARNING SKY. A dull Indian-red sky at sunset, or at sunrise, warns of rain within the following 24 hours, possibly accompanied by strong winds.

PLATE 17

Weather Forecasting with Verse and Clouds

GOOD WEATHER—Cumulus

*"When white clouds cover the
heavenly way
No rain will mar your plans
that day."*

RAIN WARNING—Altostratus

*"When moon or sun is in its
house
Likely there will be rain
without."*

SUN or RAIN—Altostratus

*"Rainbow in the morning
Travelers take warning.
Rainbow at night
Travelers' delight."*

GOOD WEATHER—Stratocumulus

RAIN WARNING—Stratocumulus

*"...When it is evening...it wil
be fair weather for the sky is red
And in the morning it will be fou
weather today for the sky is re
and lowering."* Matt. 16:2–

RAIN WARNING—Cirrostratus

RAIN and SQUALL—Stratus

*"Last night the moon had a
golden glow
Tonight no moon I see."*
Longfellou

RAIN WARNING—Stratocumulus

*"Evening gray and morning red
Sends the traveler wet to bed
Evening red and morning gray
Sends the traveler on his way."*

PLATE 17

UNCERTAIN SKY—Cirrus above 45°

GOOD WEATHER—Cirrus fibratus

GOOD WEATHER—Cirrus uncinus

*"Mackerel skies and mares' tails
Make tall ships carry low sails."*

RAIN WARNING—Altostratus

*"If smoke and birds go high
There's no rain in the sky.
If smoke and birds are low
Watch out for a blow."*

RAIN WARNING—Cumulus congestus

GOOD WEATHER—Altostratus

RAIN and SQUALL—Cumulonimbus

*'In the morning mountains . . .
'n the evening fountains."*

Weather forecasts are only a recent invention. From the dawn of time until a hundred years or so ago, men relied on watching the clouds and memorizing verses to make their predictions of what tomorrow's weather will be.

They knew that certain atmospheric conditions were likely to produce certain kinds of weather and they put that knowledge into verses or proverbs, which passed on from generation to generation, and modern meteorological discoveries have since confirmed the accuracy of the old sayings.

PLATE 18

L9—NIMBOSTRATUS

L9—CUMULONIMBUS capillatus

L5—STRATOCUMULUS opacus

L3—CUMULONIMBUS calvus

H6—CIRROSTRATUS covering sky

H7—CIRROSTRATUS with halo

L9—CUMULONIMBUS

MI—ALTOSTRATUS translucidus

will continue fair—because this condition exists only when the atmosphere is clear and devoid of moisture.

But when the moon appears to us dull and pale, cirrostratus clouds are forming, and other weather-making clouds are on the march. This is not a sign of immediate rain, only preparation for it.

A small ring around the moon has no particular meaning. This condition only tells us that some very thin cirrostratus clouds are in the air. But when we see a large ring, or halo, around the moon, with the winds from northeast to the southern points, we can be certain that unsettled weather is not far away.

If the halo moon then changes to altostratus clouds—when the moon seems to be behind frosted glass—the chances are that some precipitation will occur within 12 to 24 hours, if the prevailing winds are from northeast to south. These same cloud conditions, but with other winds, may produce an overcast sky, but no rain. The altostratus clouds usually travel from 15 to 20 hours ahead of the stratocumulus and nimbostratus clouds which bring the actual precipitation.

These particular cloud formations—when the moon is shining—enable the forecaster to make accurate observations at night, as well as in the daytime.

It is practically impossible to dissuade some thousands of people from the belief that seeds *must* be planted according to the sign of the moon, mainly on the "full of the moon."

What really happens is: The moon being full and bright provides the gardener with better illumination. So, he does a more thorough job of his planting! The moon has nothing whatever to do with how well, or how badly, the seeds sprout and grow, or how the plants bear the fruit expected of them.

Coincidence has fostered many firm beliefs; stubbornness has nourished them.

Bears, squirrels, and other animals do not prophesy colder winters by growing thicker coats and heavy fur, nor birds by heavier feathering. Thickness of pelts and density of feathers is due to the weather of the preceding spring and summer months. Frequent rains simply provided plenty of fruits, berries, and grains for these untamed creatures; the easy and abundant feeding put the birds and the beasts in top physical condition.

Through the ages the common cricket—familiarized in Charles Dickens' *Cricket on the Hearth*—has been considered a weather "prophet." This friendly insect *does* have a way about him that fits into our temperature. With a little patience, we can put the tiny fiddler to the test. The eminent educator and inventor Amos Emerson Dolbear had a pet cricket. At any time, this cricket could provide Dolbear with the temperature. The only requirement was that Dolbear count the number of chirrups per minute, add 100, then divide by 4. *Presto!*—Dolbear knew how hot or how cold it was in the room. For example: Counting 188 chirrups per minute, the temperature then was simple mathematics:

$$188 + 100 = 288$$
$$288 \div 4 = 72 \text{ degrees F.}$$

Another way is to count the chirrups for 14 seconds, then add 32. This will give us the "correct" temperature. We will need an alert mind, knowledge of mathematics, and a good watch. Why not just glance at the thermometer?

There is little, if any, dependence to be placed in the many freak weather gadgets such as a mule's tail wagging, a figure coming out of a little house, or a girl's dress changing its color. For such things to indicate rain they must be dripping wet anyway—so should we go outdoors to "read" the gadget we'd become drenched, making of our ownself a weather gadget! A blotter will soak up moisture from the air, indicating what is going on at the time in the atmosphere, but not foretelling.

"If it rains on St. Swithin's Day, it will rain for 40 days." This bit of weather folklore must have originated on a St. Swithin's Day, somewhere when it did actually rain and continue raining for 40 days. It is, of course, mere superstition.

And . . . *the poor, poor maligned and celebrated groundhog!* Surely, if the groundhog comes out of his darkened den and the sun is shining brightly, the abrupt blazing of the light hurts his eye! Wouldn't it hurt ours? So the groundhog turns and darts back into his hole. The groundhog can foretell the weather no better than can the whangdoodle, and the whangdoodle has been pining away for years.

A cloudburst is not a cloudburst. Clouds cannot and do not *burst.* At times, the precipitation from a cloud is so heavy as to cause a person to think the heavens have been split asunder.

A waterspout is not a waterspout. So called, this phenomenon is created by the same conditions that produce a tornado, but over water. Although the "waterspout" funnel, from cloud to ocean surface, can be hundreds, even thousands, of feet long, the actual water, or water spray, involved would hardly be more than 6 to 10 feet high, and its maximum is seldom over 32 feet above the ocean surface. Sailors believed that firing a cannon would break up a waterspout. Any "success" was purely coincidental, the gunfire having no effect on the weather-born monster. Like the tornado, the so-called waterspout is very brief in duration.

The sun does not "draw" water. We are only seeing the sun shining between clouds and the dust in the air. The technical name of these "rays" is *crepuscular.*

Black frost is the same as white frost, except that black frost is white frost with a glaze, usually a killing frost. The glaze, being transparent, causes the earth beneath to appear dark, or black.

Clouds appear to be dark because they are withholding their moisture and reflecting the colors of the earth, rather than the light from the sun.

As to storms: *"Long indicated, long to last; short notice, soon to pass."* Thunderstorms come in a hurry, leave in haste. Since thunderstorms arrive ahead of a cold front, they are of short duration. But if the weather has been cloudy, torpid, humid, and threatening for several days, when the precipitation does come it will usually be of long duration—often coming in with the warm fronts, which can last for several days.

"Red sky at night, sailor's delight; red sky in the morning,

sailors take warning." This can apply as well to landlubbers. The fact is, this is true most of the time. A red sky in the morning will be seen in the western sky, indicating by the clouds that make the red sky possible that wet weather is on its way to us from the east. Therefore, "take warning." The evening red sky is to the east, when the sun is, of course, in the west, indicating that the rain has passed or is moving out (see Plate 17).

"*If fleecy white clouds cover the heavenly way, no rain should mar your plans that day.*" Sound folklore. Fleecy white clouds are cumulus humilus clouds, good weather clouds. The high icy cirrus clouds are also good weather clouds. If precipitation were near, these particular cloud formations would not be visible to us.

"*Frost or dew in the morning light, shows no rain before the night.*" For either frost or dew to form (dew does not fall, it forms, as does "sweat" on the outside of a glass of cold water, by water vapor condensing on cooler objects), the air must be clear and cool. Neither of these phenomena is the result of weather-making cloud formations. This folklore gem is based upon logical weather forecasting.

The "woolly bear" caterpillar is believed to foretell "warmer" winter if the larva's center band is wide and dark. But if this band is narrow, the coming winter will be colder than normal. Actually, intrigued scientists have observed this closely. Surprisingly, the "forecast" has proved true more often than not. But a rational person would put this in the column of coincidence. If insects functioned by reasoning powers, rather than by instinct alone, then we would be tempted to change our view.

"*Mackerel sky and mares' tails, make lofty ships carry low sails.*" The mackerel sky is composed of cirrus and cirrocumulus clouds. They appear hours before changing weather and usually indicate strong winds ahead of the front moving in. The stronger the wind, the less sail would be carried for safe navigation in a rough sea.

"*Rain before seven, clear by eleven.*" Positively wrong as

many times as right. Since the average shower is of but a few hours' duration, it follows that if rain is falling in the early morning, by 11 o'clock the weather will clear. However, a warm front often will bring rain of longer duration. It is not unusual for rain beginning at dawn to continue for several days and nights—ignoring the 11 A.M. deadline.

"Ring around the moon, the sign of rain; larger the ring, nearer the rain." This sign has a fifty percent chance of being accurate—because a ring around the moon indicates rain is coming soon. Without moisture in the air we would see no ring. But the chances of rain soon are better if, instead of the ring, we see the halo, the presentation of which requires much more moisture in the air.

"Mountains in the morning, fountains in the evening." This applies to mountains of clouds in the sky. High, billowing cumulus clouds may build into cumulus congestus clouds in midmorning, to develop into cumulonimbus thunderstorm clouds by late afternoon (see Plate 17).

"If freezing into April, the barn you will fill; if green you see in January, fill and lock your granary." If the weather is warm in January, an early thaw is likely. This activity affects the condition of the soil, and thus vegetables and early grains could become damaged by the cold that follows in February and March. If we have continuous cold winter until April, then the ground would not be disturbed by an early thawing-freezing-thawing-freezing. More likely, we would have a gradual thawing. Consequently, better vegetables and grain production results.

"Short notice, soon to pass; long notice, long will last." This folklore observation is accurate. It is one we find dependable in forecasting from cloud formations and cloud sequences.

Flying birds can give indications of the weather to be. When humidity is high (much humidity), birds fly lower to the ground. This foretells a change in the weather, but not necessarily rain. This could just as well mean overcast. When the birds are flying higher, the air is clearer and less humid.

That mountains make their own weather is a fact. Winds forced up one side of a mountain and down the other often cause clouds, fog, and precipitation to develop on the windward side. This is why mountainous areas regularly receive more rainfall than surrounding low lying regions.

THE POSSIBLE effect of volcanic eruptions on our weather has long been a popular topic of conversation. Good old Ben Franklin was probably the first scientist to actually try to correlate a volcanic eruption with changes in the weather. He theorized that volcanic ash clouds spewed several miles up into the atmosphere could cause colder-than-normal weather over large regions of the hemisphere.

Since Franklin's day, many other scientists have also studied the volcano-weather connection, and have come to the same conclusions. The historic "biggies," such as Tambora and Krakatoa, can claim at least some responsibility for unusually cold weather over many parts of the world. More recently the eruption of El Chichón may have caused some changes in our weather, although precise effects are impossible to determine.

What, then, is this so-called volcano-weather connection?

With any volcanic eruption, large or small, certain amounts of ash and other particles are shot into the air with incredible force. If the volcano is especially powerful, some of this "smoke" and ash will rocket several miles skyward, actually reaching beyond the troposphere—into the stratosphere. The tranquil stability of this layer of the upper atmosphere can "hold" the volcanic ash cloud in virtual limbo for many months, even years. As the cloud makes its temporary home in the stratosphere, it begins to spread and smear over very large areas, bigger than some continents.

This atmospheric "blanket" may allow less of the sun's

Fig. 14 Aerial view of Mount St. Helens, Washington, in eruption, May 18, 1980. *Photo courtesy U.S. Geological Survey.*

warmth to enter the lower atmosphere for a time, thereby causing a cooling at the earth's surface. It may be insignificant—or consequential, such as the icy classic, "the year without a summer" in 1816 which followed the eruption of the mammoth Tambora.

The cooling effect from major eruptions can last for several years, but is most highly concentrated during the first one or two years immediately following the blast. The toughest part in analyzing all of this is the *pinpointing* of specific weather changes as they may be related to a volcanic eruption. It is, in fact, impossible to predict, for example, that Chicago will have a raging blizzard on January 29, or that Dallas will suffer a hard freeze on November 23. We simply cannot be that specific. In fact, even longer-term weather effects over larger areas are difficult to estimate. One *possible* effect: Rainfall may decrease slightly over large parts of the world following a major eruption, the result of a "slowed" evaporation of water from the earth's surface. Or the *jet streams*—fast-moving rivers of air imbedded in the upper level winds—may be shifted, causing changes in the paths of weather systems as they circle the globe.

Science has yet to establish the definite correlations between volcanoes and our weather. Until then, at least we can still enjoy *trying* to outguess mother nature, using this additional tool in our forecasting arsenal.

APPENDIX—Build Your Own
Weather Station

Few hobbies are as interesting, educational, and consistently attractive as weather-watching. Also, setting up a small weather station is not only inexpensive, the upkeep is practically nothing.

Why not learn, firsthand, as much as possible about the weather—today, tomorrow, next year—the subject that concerns everyone more than anything else? Once understood, the weather can be a real pleasure.

Professional forecasters employ several fundamental instruments which are necessary in preparing their daily or extended forecasts. Without these instruments, records of the weather could not be properly assembled. Such records are very valuable, for they provide comparisons of conditions from year to year—such as temperatures, rainfall, snow accumulations, wind velocity, humidity—and these factors can be of great worth to agriculture, industry, or any phase of endeavor that can benefit from knowing of past happenings that may relate to the present and to the future.

BAROMETERS

The barometer is the most important of the weather instruments. It indicates atmospheric pressure. Its reading can change within minutes, as in the case of the passing of a storm.

One type of barometer is the mercury barometer, activated by the expansion and contraction of a mercury column. This movement is produced by the rise or fall of the air pressure, indicating the approach or passage of storms.

Another type is the aneroid barometer. This device is activated by air pressure on a vacuum contained within thin metal membranes. Any variation in the air pressure causes these "elastic" membranes to respond. The movement is transmitted by attaching a spring to an indicating needle.

THERMOMETERS

Thermometers and barometers were invented many years before these instruments were fashioned to register temperature accurately. Not until 1714, when the German physicist Gabriel Daniel Fahrenheit devised a glasstube thermometer, using a column of mercury for the indicator, was this accomplished. Fahrenheit designed his reading scale to show 32° F as freezing point and 212° F as boiling point. The acceptance of his thermometer and its scaling has endured.

The centigrade thermometer, first described by the Swedish astronomer Anders Celsius in 1742, differs from the Fahrenheit thermometer in that the divisions are in one hundred parts, with zero corresponding to 32° F and 100° corresponding to 212° F.

In a measure, we can determine if the temperature has risen or fallen without consulting the thermometer by simply glancing through the window at the skies. If the sky was clear at night, but by morning has become cloudy, then we might guess that the temperature has risen overnight. This is because the forming clouds prevented the heat radiation from leaving the earth. On the other hand if we observed cloudiness at night, but by morning the sky has cleared, we can surmise that it is cooler outside than when we went to bed.

The most interesting thermometer to obtain for our weather station is the max and min thermometer. This instrument is moderate in cost. The most compact ones are the U-type with a double reading— one side showing how high the temperature went during any period, and the other side showing how low the reading went. With such a thermometer we can, with accuracy, be able to tell exactly how cold, or how hot, the air was during the night–or during the day. The lowest reading usually comes just before sunrise, not the best time to crawl from the covers to spotcheck the temperature.

For ordinary use, the small mercury-type thermometer is probably the best, since it is both accurate and inexpensive.

Another interesting thermometer is the indoor-outdoor type. It enables us to read the temperature on the outside and the inside of a building simultaneously.

WIND VANES

The simplest way to determine the wind's direction is to wet a finger, hold it aloft, then decide which side of the digit becomes cool. That side (cooled by quick evaporation from air movement) informs us of the wind direction.

An inexpensive wind vane can be made from scrap wood or metal and the instrument mounted so that it will move freely with the wind from the top of the pole, pipe, or pivot set on any roof or chimney. Low-cost attractive vanes can be readily purchased.

Most of the mechanical vanes that you can buy work from the outside, but have a "remote" readout, either digital or dial-type, that can be wired to the inside of your house.

A companion instrument for the wind vane is the anemometer. This device indicates the speed of the wind, helpful knowledge in foretelling the weather, especially when used in conjunction with the cloud formations.

RAIN GAUGES

There are many types and styles of rain gauges; the simplest and cheapest is the basic plastic cup with graduated markings on the side. This is perfectly adequate for most normal use. More elaborate, and of course expensive, gauges can have remote readout capability with precisely calibrated measuring sensors.

No matter which style rain gauge is used, measurements are usually made in hundredths of an inch. On the plastic, cylinder-type rain gauge, the markings on the side can be easily read to determine just how much rain has come down.

No fancy equipment is required to measure snowfall. Simply take out a ruler, and stick it in the snow to determine how deep it is.

Relative humidity is measured with a hygrometer or psychrometer. It determines how much water vapor is in the air, compared to the maximum amount that the air can hold at the same temperature, and expresses it as a percentage. A simple digital or dial-type hygrometer is easy to buy, and makes a useful addition to the home weather station.

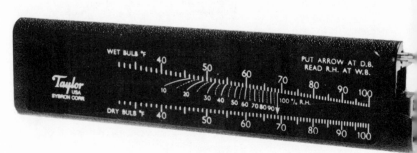

Fig. 15 Psychrometer. *Courtesy Taylor/Sybron.*

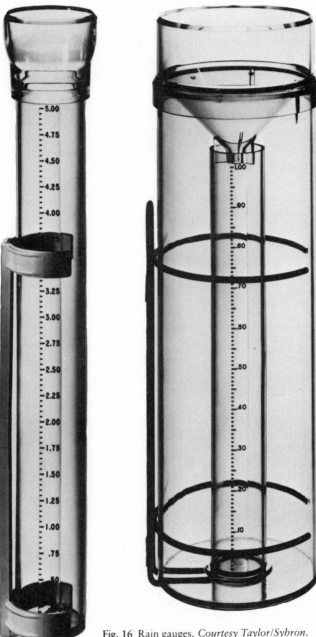

Fig. 16 Rain gauges. *Courtesy Taylor/Sybron.*

Fig. 17 Barograph. *Courtesy Taylor/Sybron.*

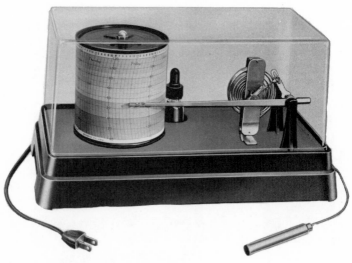

Fig. 18 Thermograph. *Courtesy Taylor/Sybron.*

Fig. 19 Weatherscope, with wind direction and wind speed indicator, barometer and maximum and minimum thermometer. *Courtesy Taylor/Sybron.*

The full-color cloud charts described in this book are available from Cloud Chart, Inc., P.O. Box 29294, Richmond, Va., 23233, as follows: 17½ × 22 chart, $3.00; 11 × 17 folklore chart, two-sided, $3.00; elementary grades chart, $2.00.